Oil Well Testing Handbook

Oil Well Testing Handbook

Editor

Suhas Kulkarni

Oil Well Testing Handbook

Edited by **Suhas Kulkarni**

Printed in 2017

ISBN: 978-1-68117-427-3

Library of Congress Control Number: 2015936541

© 2016 by

SCITUS Academics LLC,
616, Corporate Way, Suite 2, 4766,
Valley Cottage, NY 10989

www.scitusacademics.com

Contents

Preface

Oil well test analysis is a branch of reservoir engineering. Information obtained from flow and pressure transient tests about in situ reservoir conditions are important to determining the productive capacity of a reservoir. Pressure transient analysis also yields estimates of the average reservoir pressure. The reservoir engineer must have sufficient information about the condition and characteristics of reservoir/well to adequately analyze reservoir performance and to forecast future production under various modes of operation. The production engineer must know the condition of production and injection wells to persuade the best possible performance from the reservoir. Pressures are the most valuable and useful data in reservoir engineering. Directly or indirectly, they enter into all phases of reservoir engineering calculations. Therefore accurate determination of reservoir parameters is very important.

Editor

Numerical Well Testing Interpretation Model and Applications in Crossflow Double-Layer Reservoirs by Polymer Flooding

Haiyang Yu[1], Hui Guo[1], Youwei He[1], Hainan Xu[1], Lei Li[1], Tiantian Zhang[2], Bo Xian[3], Song Du[4], and Shiqing Cheng[1]

[1]MOE Key Laboratory of Petroleum Engineering, China University of Petroleum, Beijing 102249, China

[2]Department of Petroleum Engineering, University of Texas at Austin, Austin, TX 78712, USA

[3]Research Institute of Exploration and Development, PetroChina, Korla 841000, China

[4]Department of Petroleum Engineering, Texas A&M University, College Station, TX 77843, USA

ABSTRACT

This work presents numerical well testing interpretation model and analysis techniques to evaluate formation by using pressure transient data acquired with logging tools in crossflow double-layer reservoirs by polymer flooding. A well testing model is established based on rheology experiments and by considering shear, diffusion, convection, inaccessible pore volume (IPV), permeability reduction, wellbore storage effect, and skin factors. The type curves were then developed based on this model, and parameter sensitivity is analyzed. Our research shows that the type curves have five segments with different flow status: (I) wellbore storage section, (II) intermediate flow section (transient section), (III) mid-radial flow section, (IV) crossflow section (from low permeability layer to high permeability layer), and (V) systematic radial flow section. The polymer flooding field tests prove that our model can accurately determine formation parameters in crossflow double-layer reservoirs by polymer flooding. Moreover, formation damage caused by polymer flooding can also be evaluated by comparison of the interpreted permeability with initial layered permeability before polymer flooding. Comparison of the analysis of numerical solution based on flow mechanism with observed polymer flooding field test data highlights the potential for the application of this interpretation method in formation evaluation and enhanced oil recovery (EOR).

INTRODUCTION

Over the past several decades, many EOR methods were researched in laboratories and oilfields to improve oil recovery, for example, polymer flooding [1], surfactant flooding [2], alkali-surfactant-polymer (ASP) flooding [3], nanoparticles [4, 5], low salinity water flooding [6], and CO_2 [7, 8]. However, polymer flooding is most commonly applied in oilfields, especially hydrolyzed polyacrylamide (HPAM) polymer flooding because of its low cost and high efficiency [9]. The oil recovery of polymer flooding is enhanced mainly by increasing sweep efficiency [10].

Conventional pressure transient test has historically been the main application of permeability and skin estimation in oilfields, by using a pressure gauge positioned at a fixed depth in a well. The pressure test of multilayered reservoir was studied from the 1960s; however, the research on the individual production of multilayered reservoir was not carried out, due to the restriction of testing tools and technology. A percolation model of multilayered reservoir was derived in 1961, and the wellbore pressure and production of individual layers were also deduced [11]. This model considered that the interlayer had different parameters but neglected the wellbore storage effect. In 1978, a new model was further developed to get the wellbore pressure solution in real space for multilayered reservoir by using Stehfest algorithm [12]. It took the wellbore storage and skin factor into account, whereas it ignored the crossflow of wellbore pressure response. From the 1980s to 1990s, many researchers interpreted well testing data by analysis of measured wellbore pressure and stratified flow rate. With the help of multilayer testing techniques, the expression of pressure solution was established through the relationship between wellbore pressure and stratified flow rate of multilayered reservoir [13, 14]. The well testing model of crossflow double-layer reservoir was put forward in 1985 [15], which was further investigated by theoretical study of flow mechanics [16]. However, the type curves of crossflow double-layer reservoirs were not established. The problem of interlayered crossflow in a stratified reservoir was mathematically simplified by employing a semipermeable wall model [17]. Based on the former research, the dynamic model and exact solution of bottom hole pressure were proposed. Most researches on well testing and fluid percolation in double-layer reservoirs were based on analysis method to get the analytic solution of bottom hole pressure (BHP). In recent years, the numerical methods were employed to study well testing problems of multilayered reservoir with the help of rapid development of computer technology [18–21].

HPAM polymer solution is one kind of non-Newtonian fluids, and its viscosity is a significant parameter used to establish well testing interpretation model for polymer flooding. Many researches on the rheological behavior of polymer solution simply consider polymer as power law fluid and using constant power exponent model to represent the percolation of polymer solution in reservoirs [22–25], which is unable to meet the actual demands of our oilfields. For crossflow double-layer reservoirs by polymer flooding, there exist not only shear effect and viscoelastic effect but also physic-chemical interaction during polymer solution percolating in porous medium, whereas the constant power exponent viscosity model ignores diffusion and convection of polymer during transport in porous medium. Meanwhile, the adsorption of polymers in the porous medium results in IPV [1, 6,26, 27] and permeability reduction [28–31], which also needs to be taken into account.

At present, well testing models and techniques in double-layer reservoirs by water flooding become mature, and commercial software can be used for reservoir evaluation; however, well testing models and interpretation methods in reservoirs with crossflow by polymer flooding are still less. The purpose of this study is to establish well testing interpretation method that can be applied in crossflow double-layer reservoir by polymer flooding, by considering shear, diffusion, convection, IPV, permeability reduction, wellbore storage effect, and skin factors. Moreover, field test data are further interpreted by this method for formation evaluation and EOR.

POLYMER RHEOLOGY IN POROUS MEDIUM

Materials

A proprietary HPAM used for polymer flooding was supplied by CNPC. The degree of hydrolysis is 25% and molecular weight of HPAM is 4050. The formation brines used in this study were prepared with salts of NaCl, $MgCl_2$, $CaCl_2$, and Na_2SO_4, and the synthetic brine composition is listed in Table 1. The total salinity, the sum of the ionic concentration, is 4.3 wt% (43000 ppm or 42.95 g/L).

Table 1: Synthetic brine composition

Total salinity	NaCl	MgCl2	CaCl2	Na2SO4
4.3 wt%	3.44 wt%	0.18 wt%	0.64 wt%	0.04 wt%

Rheological Model

Polymer solution was assumed to behave as pseudoplastic non-Newtonian fluid. As discussed above, the power law model [32] or Carreau model [33] cannot accurately illustrate rheological behavior of the polymer used in our case. In this study, polymer shear-thinning behavior was simulated by use of Meter equation [34]:

$$\mu_p = \mu_\infty + \frac{\mu_p^0 - \mu_\infty}{1 + (\gamma/\gamma_{1/2})^{P_a-1}} = \left(\mu_w + \frac{\mu_p^0 - \mu_w}{1 + (\gamma/\gamma_{1/2})^{P_a-1}} \right), \tag{1}$$

where μ_p is apparent viscosity of polymer solution; μ_∞ is viscosity of polymer solution at infinite shear rate, which is simplified as brine viscosity (μ_w) and satisfied the accuracy in this study since polymer concentration is relatively low and its viscosity at infinite shear rate is pretty close to brine viscosity; $\gamma_{1/2}$ is the shear rate at which apparent viscosity is the average of μ_∞ and μ_p^0; γ is the effective shear rate; P_a is a fitting parameter (usually $1.0 < P_a < 1.8$); μ_p^0 is the viscosity at very low shear rate, which is calculated by modified Flory- Huggins equation [35]:

$$\mu_p^0 = \mu_w \left[1 + \left(A_1 C_p + A_2 C_p^2 + A_3 C_p^3 \right) C_{SEP}^{SP} \right], \tag{2}$$

Where A_1, A_2, and A_3 are fitting parameters obtained from matching experimental data; C_p is polymer concentration; C_{SEP}^{SP} represents the effect of salinity and hardness on polymer viscosity.

Since temperature significantly affects rheological behavior of polymer and the effect of pressure on polymer viscosity is negligible compared with temperature, the polymer solutions were prepared by mechanical stirring at 75°C to simulate reservoir temperature. The tested polymer concentrations range from 100 mg/L (0.1 g/L or 0.01 wt%) to 4000 mg/L (the polymer concentrations in our field tests are between 1600 mg/L and 2500 mg/L). The polymer rheological measurement was carried out by Haake RS6000 rheometer made in Germany.

The viscosity of polymer solutions with different concentrations was measured at 75°C to get the fitting numbers of A_1, A_2, and A_3, shown in Figure 1 and Table 2. The measurements were performed under 0.01 s^{-1} shear rate, since μ_c is the viscosity at very low shear rate.

Table 2: Characteristics of polymer solutions

μw,(mPa·s)	A1, (g/L)−1	A2, (g/L)−2	A3, (g/L)−3	Cp0, (g/L)	D, (cm2/s)
0.5	0.642	0.201	0.931	1.750	0.0246

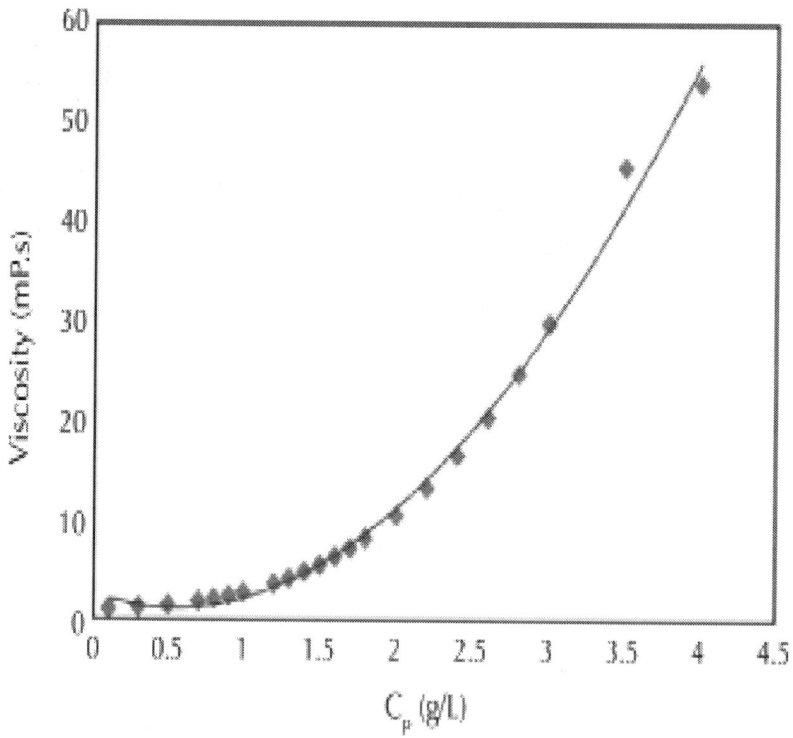

Figure 1: Relationship between polymer viscosity (μ_p^0) and polymer concentration (C_p) at 75°C under 0.01 s^{-1} shear rate.

P_a and $\gamma_{1/2}$ are functions of μ_p^0 (or polymer concentration); the expressions are provided by CNPC based on their former research, shown in the following equations, respectively:

$$P_a = 1.182\left(\mu_p^0\right)^{0.0341},$$

(3)

$$\gamma_{1/2} = 376.2\left(\mu_p^0\right)^{-1.365} + 0.0341.$$

(4)

The relationship between effective shear rate γ and seepage velocity is shown in the following [36]:

$$\gamma = \frac{3n+1}{n+1}\frac{10^4 v}{\sqrt{8C'K\phi}},$$

(5)

$$v = \frac{Q}{2\pi rh},$$

(6)

Where n is the bulk power law index, in the range of 0 to 1; C' is tortuosity coefficient; ϕ is porosity; K is permeability; Q is flowrateof injectedpolymer solution; h is reservoir thickness; r is radial distance; V is Darcy velocity. By considering IPV and permeability reduction caused by polymer flooding, (5) is changed to

$$\gamma = \frac{3n + 1}{n + 1} \frac{10^4 v}{\sqrt{8C'K_p\phi_p}},$$

(7)

Where K_p is effective permeability, $K_p = K/R_{k'}$, Rk being permeability reduction coefficient; ϕ_p is effective porosity, $\phi_p = (1 - IPV)$. During transport in porousmedium, polymer concentration is also affected by convection and diffusion. Thus, polymer concentration by considering convection and diffusion is shown in the following [37]

$$C_p(r,t) = \frac{C_{p0}}{2} - \frac{C_{p0}}{2} \, \text{erf} \left[\frac{r - Vt}{2\sqrt{Dt}} \right],$$

(8)

Where C_{p0} is initial polymer concentration; D is diffusion coefficient.

There are several shear-thinning rheological models developed for polymer solutions. The model used in this study can accurately match the apparent viscosity of the proprietary HPAM polymer provided by CNPC over a wide range of injected velocity, especially when polymer solutions pass through the perforation.

WELL TESTING MODELING METHOD-OLOGY

The percolation of polymer flooding in crossflow double-layer reservoir is sketched in Figure 2. Crossflow occurs in the interlayer and fluids can transport from low permeability zone to high permeability zone when polymer solutions are injected into the reservoir. The hypotheses are as follows: (1) polymer solutions and reservoir brines are miscible; (2) properties of polymer solutions are the same in each layer; (3) fluids flow satisfies Darcy's law; (4) each layer is homogeneous, but formation properties, for example, layer thickness, permeability, skin factor, and compressibility, are different between two layers; (5) gravity effect is negligible; (6) the initial pressure of each layer is the same, p_i; (7) reservoir rocks and fluids are compressible; (8) process of polymer transportation is isothermal; (9) crossflow of interlayer is pseudosteady state.

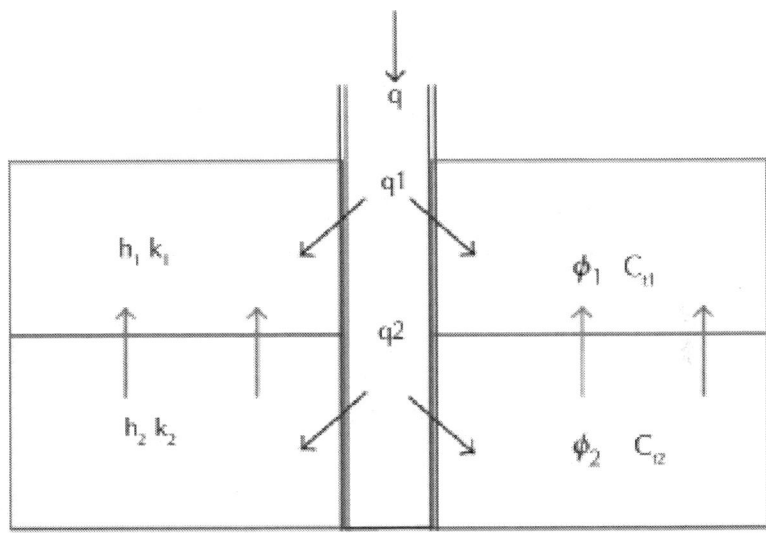

Figure 2: Sketch of polymer flooding in crossflow double-layer reservoir.

Based on the rheological model and hypotheses discussed above, the well testing interpretation model in crossflow double-layer reservoir by polymer flooding is established, by considering shear, diffusion, convection, IPV, permeability reduction, wellbore storage effect, and different layered skin factors:

1.   percolation equation:

$$K_{p1}h_1\frac{\partial}{\partial r}\left(r\frac{1}{\mu_p}\frac{\partial p_1}{\partial r}\right) + a\frac{K_{p2}h_2}{\mu_p}(p_2 - p_1) = \phi_{p1}C_{t1}h_1\frac{\partial p_1}{\partial t}$$

$$K_{p2}h_2\frac{\partial}{\partial r}\left(r\frac{1}{\mu_p}\frac{\partial p_2}{\partial r}\right) - a\frac{K_{p2}h_2}{\mu_p}(p_2 - p_1) = \phi_{p2}C_{t2}h_2\frac{\partial p_2}{\partial t};$$

$$(9)$$

2.   internal boundary conditions:

wellbore storage effect

$$qB = C\frac{dp_{wf}}{dt} - \left(\frac{K_{p1}h_1}{\mu_p}r\frac{\partial p_1}{\partial r} + \frac{K_{p2}h_2}{\mu_p}r\frac{\partial p_2}{\partial r}\right)\Bigg|_{r=r_w}$$

skin factor

$$p_w(t) = \left(p_1 - s_1 r\frac{\partial p_1}{\partial r}\right)\Bigg|_{r=r_w} = \left(p_2 - s_2 r\frac{\partial p_2}{\partial r}\right)\Bigg|_{r=r_w} ; \tag{10}$$

3.   . external boundary condition (infinite boundary):

$$p_1(\infty, t) = p_2(\infty, t) = p_i; \tag{11}$$

4.   initial condition:

$$p_1(r, 0) = p_2(r, 0) = p_i, \tag{12}$$

Where p_1 and p_2 are reservoir pressure of each layer; K_{p1} and K_{p2} are effective layered permeability; h_1 and h_2 are layer thickness; C_{t1} and C_{t2} are total layered compressibility; ϕ_{p1} and ϕ_{p2} are porosity; C is wellbore storage coefficient; s_1 and s_2 are skin factor; p_{wf} is BHP; p_i is initial reservoir pressure; a is flowrate exchange coefficient.

Dimensionless parameters are involved after solving the model and obtaining BHP:

$$p_{wDj} = \frac{\sum_{j=1}^{2}(k_p h)_j}{1.842 \times 10^{-3} q\mu_p B}(p_{wfj} - p_i), \quad (j = 1, 2),$$

$$t_D = \frac{3.6\sum_{j=1}^{2}(k_p h)_j}{\sum_{j=1}^{2}(\phi_p C_t h)_j \mu_p r_w^2}t,$$

$$C_D = \frac{C}{2\pi r_w^2 \sum_{j=1}^{2}(\phi_p C_t h)_j},$$

(13)

Where p_{wD} is dimensionless BHP; t_D is dimensionless time; C_D is dimensionless wellbore storage coefficient; μ_p is the viscosity of the first grid, which expresses the rheology behavior of fluid near wellbore.

Three new parameters are then proposed in order to effectively analyze parameters sensitivity and interpret field test data:

$$\chi = \frac{K_{p1}h_1}{K_{p1}h_1 + K_{p2}h_2},$$

$$\omega = \frac{\phi_{p1}C_{t1}h_1}{\phi_{p1}C_{t1}h_1 + \phi_{p2}C_{t2}h_2},$$

$$\lambda = \frac{ar_w^2 K_{p2}h_2}{K_{p1}h_1 + K_{p2}h_2},$$

(14)

Where χ is formation coefficient ratio; ω is storativity ratio; λ is interporosity flow coefficient.

TYPE CURVES AND SENSITIVITY ANALYSIS

Based on dimensionless BHP and dimensionless BHP derivative, the type curves of pressure and pressure derivative in log-log scale are obtained. Sensitivity analysis is further investigated.

Type Curves

Type curves of well testing in crossflow double-layer reservoir by polymer flooding are shown in Figure 3, which shows that type curves have five flow segments: (I) wellbore storage section, where pressure and pressure derivative curves are superposed, reflecting the pressure response characteristics during well storage stage; (II) intermediate flow section (transient section), that describes the pressure response from pure wellbore storage stage to mid-radial flow stage within internal region, and there is a "convexity"; (III) mid-radial flow section, where fluids flow of individual layer achieves plane radial flow before crossflow happens, showing a horizontal period of pressure derivative line; (IV) crossflow section, where fluids in low permeability layer transport through interlayer into high permeability layer, and there is a "concave"; and (V) systematic radial flow section, where the whole system presents plane radial flow over time and the pressure curve lightly turns upward due to the influence of the non-Newtonian fluid properties of polymer solution.

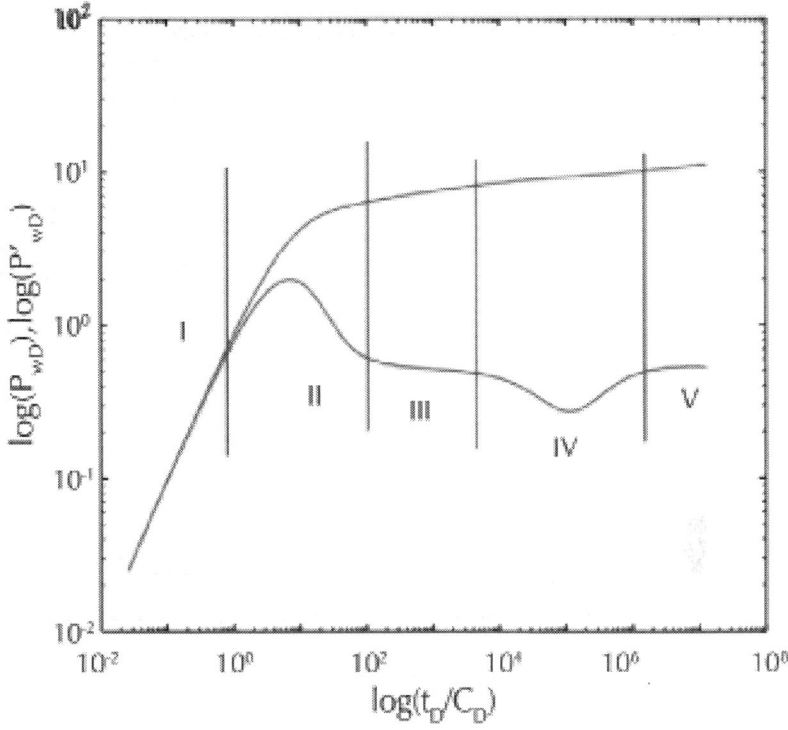

Figure 3: Type curves of well testing in crossflow double-layer reservoir by polymer flooding.

The comparison of type curves in double-layer reservoir by polymer flooding with and without crossflow is demonstrated in Figure 4. It is obvious that there exists a "concave" (section IV) in the crossflow double-layer reservoir, which is formed by the fluids percolation from low permeability layer into high permeability layer resulting in crossflow through the interlayer. After crossflow is developed over time, the "concave" will vanish and curves will overlap when pressures of each layer achieve equilibrium. In systematic radial flow section (V), the BHP in crossflow reservoir is lower than that in noncrossflow reservoir since crossflow reduces the flow resistance (equal to systematic permeability enhanced); however, the BHP derivative is the same with value of 0.5.

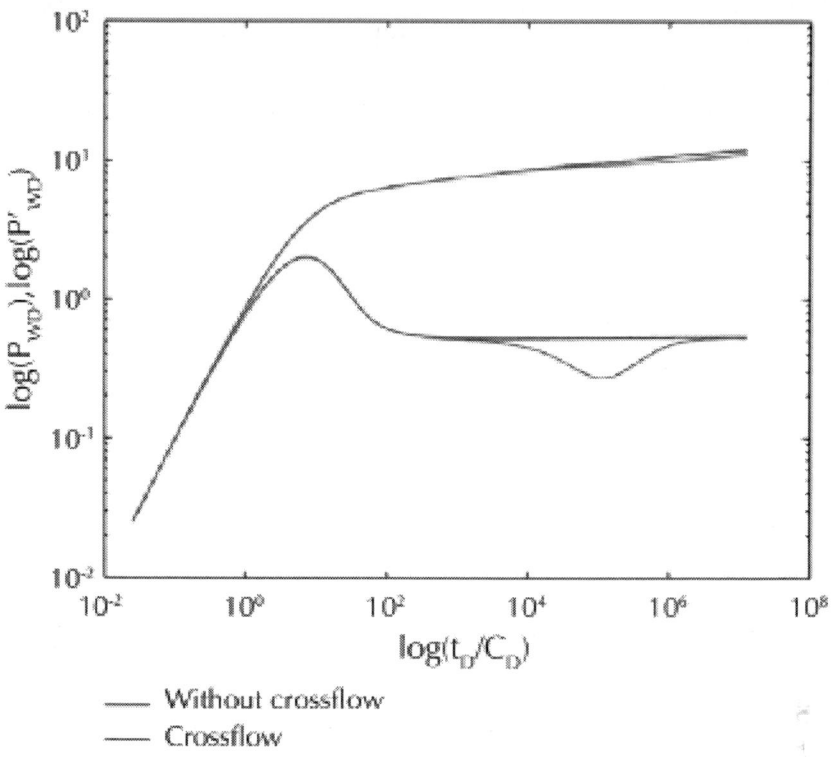

Figure 4: Type curves of well testing in double-layer reservoir by polymer flooding with and without crossflow.

Sensitivity Analysis

The effects of different parameters on type curves are investigated, including interporosity flow coefficient, ratio of formation coefficient, storativity ratio, initial polymer concentration, and IPV.

Interporosity Flow Coefficient

The influence of interporosity flow coefficient (λ) on type curves in crossflow double-layer reservoir by polymer flooding is shown in Figure 5. Smaller λ indicates fewer fluids transport through interlayer, which depends on the permeability difference and BHP difference between two layers. Smaller permeability difference or BHP difference results in small a and λ. The "concave" appears delayed with smaller interporosity flow coefficient since it needs more time for the fluids in crossflow section (IV) to achieve equilibrium. After that, individual layer reaches the plane radial flow and BHP derivative curve changes to horizontal, indicating fluids flow achieves systematic radial flow section (V). The time of "concave" appearance can qualitatively evaluate formation heterogeneity since it is influenced by layered permeability difference: it appears earlier in heterogeneous formation, and it appears later in relative homogenous formation.

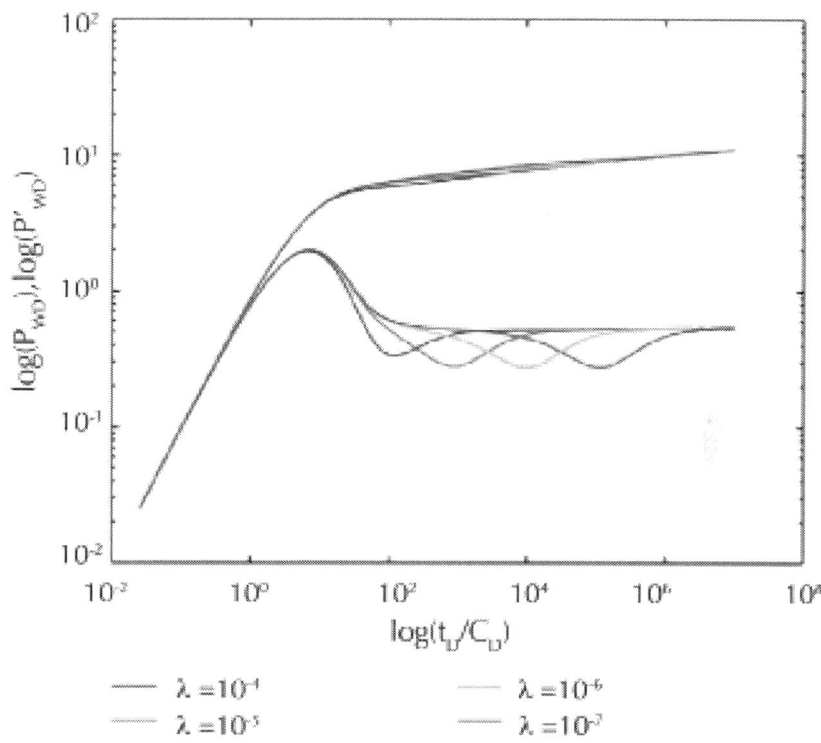

Figure 5: Effect of interporosity flow coefficient (λ) on type curves.

Formation Coefficient Ratio

Figure 6 represents the effect of formation coefficient ratio (x) on type curves in crossflow double-layer reservoir by polymer flooding. It shows that x only affects the crossflow section (IV): the smaller x is, the shallower the "concave" becomes and vice versa. For reservoirs with fixed value of layer permeability, smaller means smaller difference of layer thickness, and the "concave" becomes shallower as permeability difference decreases.

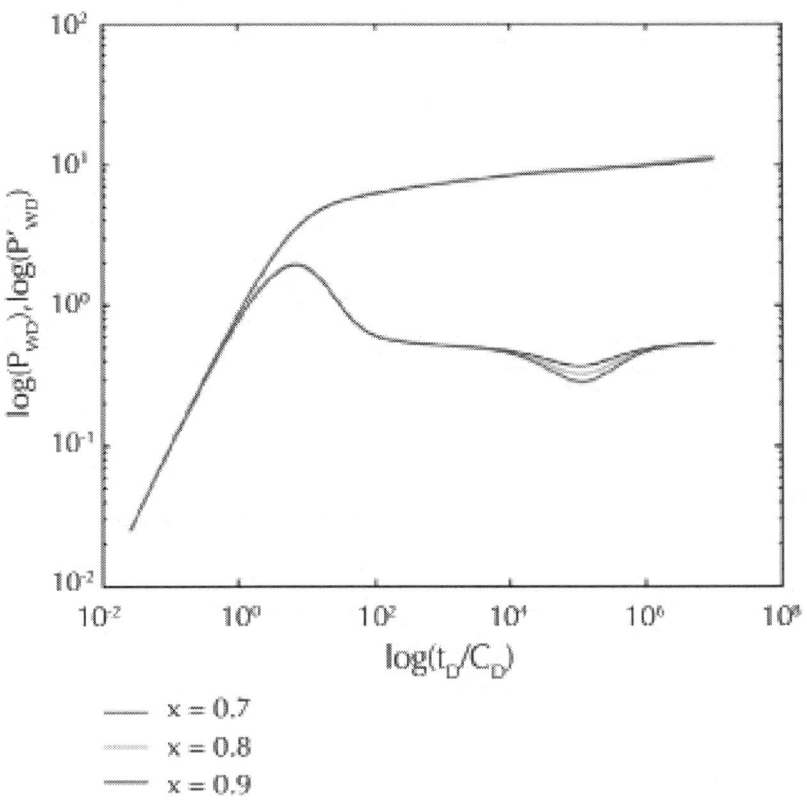

Figure 6: Effect of formation coefficient ratio (x) on type curves.

Storativity Ratio

The effect of storativity ratio (ω) on type curves in crossflow double-layer reservoir by polymer flooding is shown in Figure 7. The width and depth of the "concave" are influenced by ω the "concave" gradually becomes narrower and shallower when ω increases, and vice versa. Individual layers, respectively, reach their radial flow after the crossflow segment ends, indicating systematic radial flow section (V).

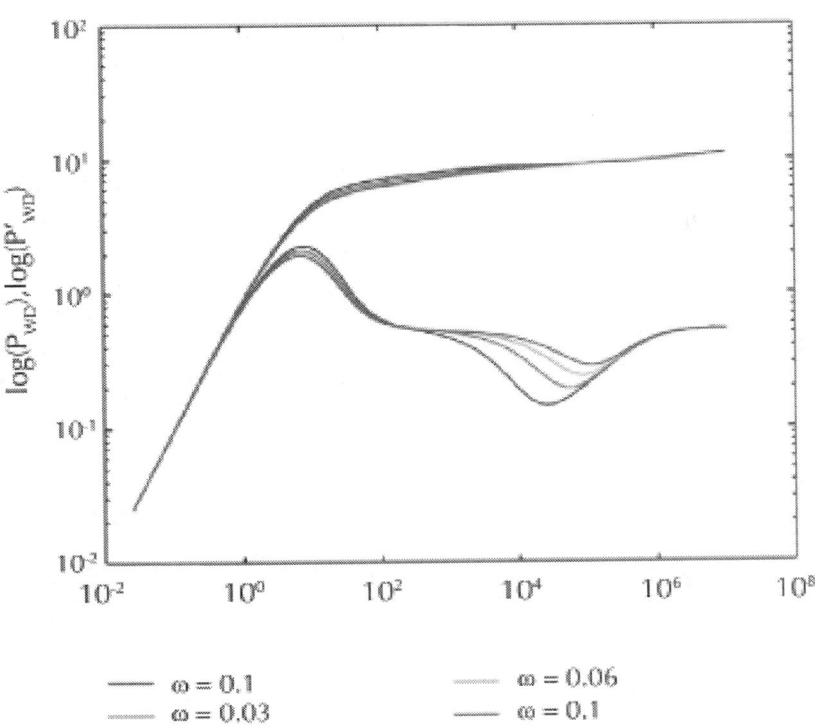

—— $\omega = 0.1$	······ $\omega = 0.06$
······ $\omega = 0.03$	—— $\omega = 0.1$

Figure 7: Effect of storativity ratio (ω) on type curves.

Initial Polymer Concentration

The effect of initial polymer concentration (C_{p0}) on type curves in crossflow double-layer reservoir by polymer flooding is shown in

Figure 8, which indicate that the crossflow section (IV) appears later and BHP derivative curve in systematic radial flow section (V) turns more upward by increasing C_{p0}. Since viscosity is increased for higher C_{p0}, there is more flow resistance for fluids to transport through interlayer, resulting in delay of crossflow section (IV) appearance and greater amplitude of BHP derivative curve in systematic radial flow section (V). Consider $C_0 = 0$ mg/L expressed as water flooding, which is Newtonian fluid with constant viscosity. Further investigation indicates that the effect of polymer rheology on type curve section (V) is dramatically reduced by crossflow, which means the pressure curve and pressure derivative curve of polymer flooding are similar to those of water flooding in section (V) and this phenomenon is also proved by field test data. However, the slope of type curves in one-layer reservoir with homogenous thickness by polymer flooding is much larger than that of water flooding.

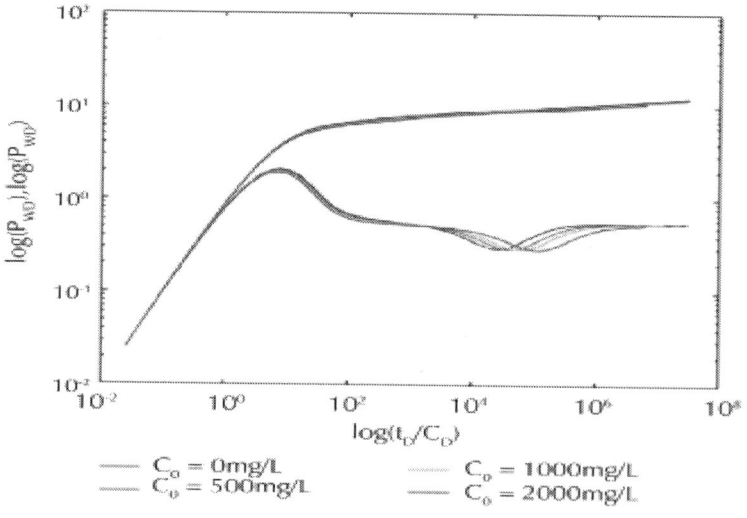

Figure 8: Effect of initial polymer concentration (C_{p0}) on type curves.

Inaccessible Pore Volume

Figure 9 represents the effect of IPV on type curves in crossflow double-layer reservoir by polymer flooding. The crossflow section (IV) appears earlier for reservoir with bigger IPV. Bigger IPV means lower effective porosity, and the fluid velocity is higher for the reservoir with fixed flow rate of polymer injected, resulting in earlier appearance of the crossflow section (IV) and systematic radial flow section (V). However, the effect of IPV on well testing type curves is unremarkable; moreover, the IPV caused by polymer flooding in oilfields is usually less than 0.15, so the effect of IPV can be negligible during well testing interpretation. Unlike other parameters, the effect of IPV on type curves is listed here only for theoretical analysis.

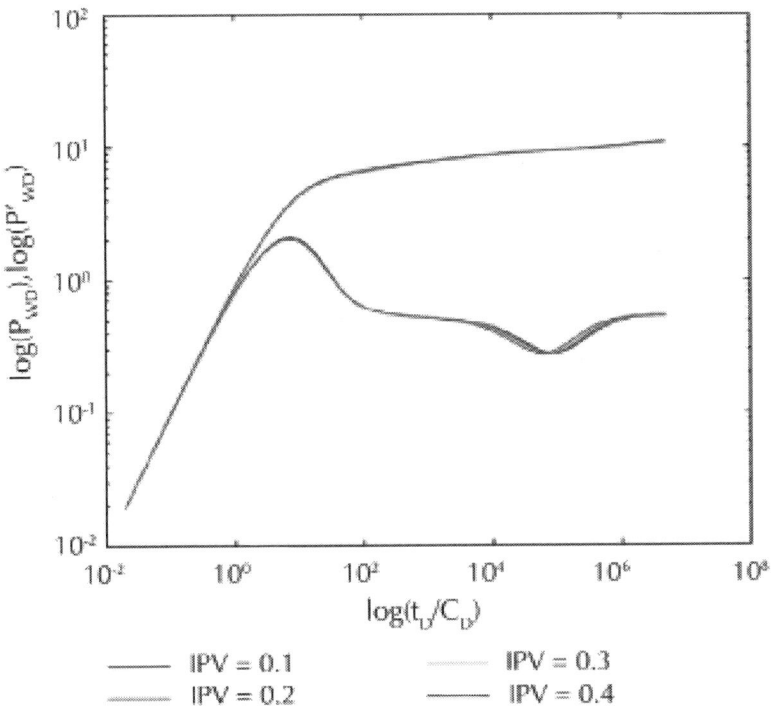

Figure 9: Effect of IPV on type curves.

Wellbore Storage Coefficient

The effect of wellbore storage coefficient on type curves in crossflow double-layer reservoir by polymer flooding is shown in Figure 10. The depth of the "concave" and "convexity" is influenced by C; however it does not affect the width. The crossflow section (IV) and systematic radial flow section (V) gradually appear earlier with ω increases; meanwhile, the mid-radial flow section (III) is shortened.

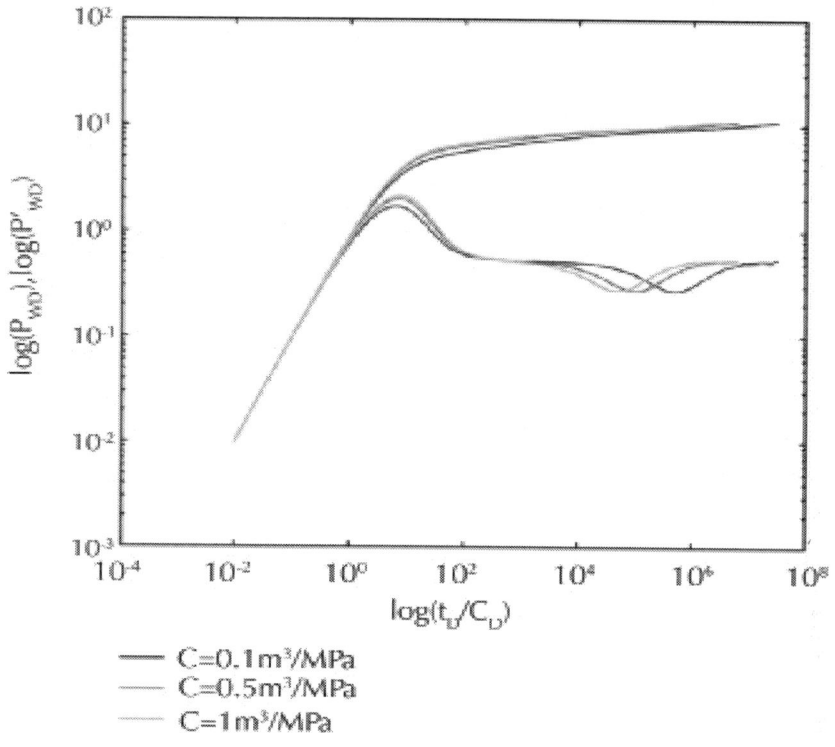

Figure 10: Effect of wellbore storage coefficient (C) on type curves.

FIELD TESTS INTERPRETATION

Well testing data of field test was provided by CNPC. Then draw the BHP data with time in log-log scale. Interpret the data and perform history matching of type curves to evaluate reservoir formation and calculate the average formation pressure, layered permeability, layered skin factor, and wellbore storage coefficient. The interpretation results of layered permeability and layered skin factor are significant for oilfields, since oil industry will adjust development plan of production based on them. If the layered permeability is much lower or the layered skin factor is much higher than those of before polymer flooding, it indicates that polymer flooding leads to serious formation damage and specific methods should be employed to reduce formation damage and improve production, for example, acidizing.

Basic Properties of Oilfield

The tectonic surface area is 33 km², and structure amplitude is about 100 m. The formation conditions and fluid properties are suitable for polymer flooding; meanwhile, relatively low salinity and low divalent cation concentration are beneficial to maintaining systematic viscoelasticity. The characteristics of crude oil under surface conditions and reservoir conditions are shown in Tables 3 and 4, respectively. The pressure derivative curve of field test data was modified for curve smoothing by using Bourdet's method [38].

Table 3: Characteristics of crude oil under surface conditions

Density (g/cm3, 20°C)	Viscosity (mPa·s, 20°C)	Viscosity (mPa·s, 55°C)
0.925~0.934	407.5~533.6	49.56~58.21

Table 4: Characteristics of crude oil under reservoir conditions

Density (g/cm3)	Viscosity (mPa·s)	Volume factor	Saturation pressure (MPa)	Oil-gas ratio	Acid number
0.8675	14.2	1.1038	12.70	42	0.4~1.16

Field Test One

Well testing was based on injection fall-off process. The polymer solutions were injected into double-layer reservoirs with initial concentration of 1600 mg/L, and the reservoir thickness is 14 m. Well 5-227 performed polymer flooding from Feb 1, 2012, to May 7, 2012, and then the polymer injection was stopped and pressures were measured. It took three days for well testing, and polymer flooding was performed again since May 10, 2012. Basic parameters of well and reservoir are shown in Table 5.

Table 5: Basic parameters of well and reservoir for 5-227 field test

Injection rate	q (m3/d)	100
Layer 1 thickness	h1 (m)	8
Layer 2 thickness	h2 (m)	6
Oil volume factor	B0	1.1037
Porosity	ϕ	0.3
Crude oil viscosity	μ0 (mPa·s)	14.2
Brine viscosity	μw (mPa·s)	0.5
Temperature	°C	75
Total compressibility	Ct (1/MPa)	0.0014
Well radius	rw (m)	0.1
Layer 1 permeability before polymer flooding	mD	1592
Layer 2 permeability before polymer flooding	mD	1466
Layer 1 skin factor before polymer flooding	n/a	1.11
Layer 2 skin factor before polymer flooding	n/a	1.18

The history matching curves and field testing data are shown in Figure 11, and the interpretation results are shown in Table 6. The permeability and skin factor of individual layer acquired by interpreting field test data are consistent with the actual situation of oilfield, indicating that our model can accurately interpret Field Test One and evaluate formation. Meanwhile, polymer flooding results in negligible permeability reduction or formation damage in this case, since the interpreted permeability and skin factors are nearly the same as those of before polymer flooding.

Table 6: Interpretation results of Field Test One (Well 5-227)

Average reservoir pressure	MPa	17.26
Layer 1 permeability	mD	1570
Layer 2 permeability	mD	1460
Layer 1 skin factor	n/a	1.13
Layer 2 skin factor	n/a	1.20
Wellbore storage coefficient	m3/MPa	0.60

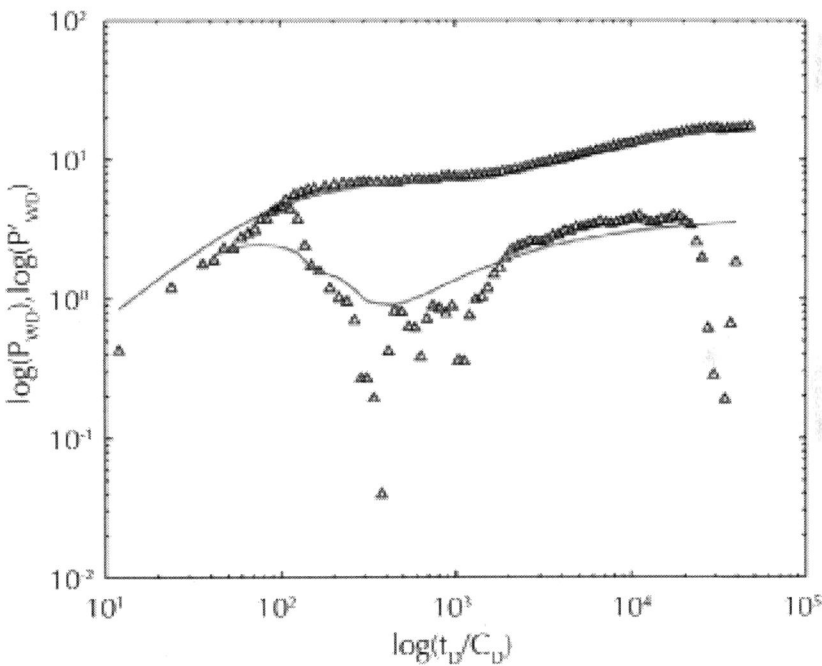

Figure 11: Field test data and history matching of type curves (Field Test One: Well 5-227).

Field Test Two

Well testing was also based on injection fall-off process. The polymer solutions were injected into double-layer reservoirs with initial concentration of 1600 mg/L, and the reservoir thickness is 21 m. Well 5-225 (500 meters away from Well 5-227) performed polymer flooding from Feb 1, 2012, to Apr. 28, 2012, and then the polymer injection was stopped and pressures were measured (nine days before Field Test One). It took three days for well testing, and polymer flooding was performed again since May 1, 2012. Basic parameters of well and reservoir are shown in Table 7.

Table 7: Basic parameters of well and reservoir for 5-225 field test

Injection rate	q (m3/d)	136
Layer 1 thickness	h1 (m)	12
Layer 2 thickness	h2 (m)	9
Oil volume factor	B0	1.1037
Porosity	ϕ	0.25
Crude oil viscosity	μ0 (mPa·s)	14.2
Brine viscosity	μw (mPa·s)	0.5
Temperature	°C	75
Total compressibility	Ct (1/MPa)	0.0014
Well radius	rw (m)	0.1
Layer 1 permeability before polymer flooding	mD	1352
Layer 2 permeability before polymer flooding	mD	211
Layer 1 skin factor before polymer flooding	n/a	2.49
Layer 2 skin factor before polymer flooding	n/a	0.37

The history matching curves and field testing data are shown in Figure 12, and the interpretation results are shown in Table 8. The occurrence of "concave" is earlier than Field Test One, due to the bigger permeability difference between two layers. The skin factor of individual layer and layer 1 permeability acquired by interpreting field test data are consistent with the actual situation of oilfield, which further prove that our model can accurately interpret Field Test Two and evaluate formation. Moreover, the layer 2 permeability is 68 mD and permeability reduction coefficient is 3.1 on average, indicating formation was damaged by polymer flooding. Blockage removal agent was further injected into the reservoir and layer 2 permeability was increased to 174 mD, resulting in 2.4% EOR of individual well.

Table 8: Interpretation results of Field Test Two (Well 5-225)

Average reservoir pressure	MPa	18.56
Layer 1 permeability	mD	1340
Layer 2 permeability	mD	68
Layer 1 skin factor	n/a	2.56
Layer 2 skin factor	n/a	1.98
Wellbore storage coefficient	m 3 / MPa	0.54

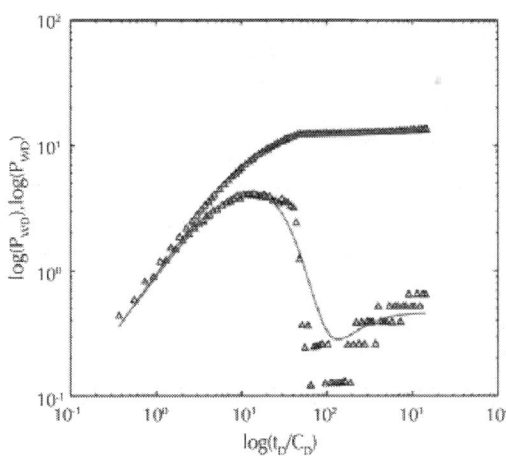

Figure 12: Field test data and history matching of type curves (Field Test Two: Well 5-225).

CONCLUSION

This work established well testing models for crossflow double-layer reservoirs by polymer flooding. Type curves of numerical well testing were obtained, and field test data were further interpreted and history-matched. The main conclusions drawn from this study are as follows.

- The model developed in this work by considering IPV, permeability reduction, shear rate, diffusion, and convection can accurately demonstrate rheological behavior of the proprietary HPAM polymer provided by CNPC over a wide range of injected velocity, especially when polymer solutions pass through the perforation.

- Type curves have five sections with different flow status: (I) wellbore storage section, where pressure and pressure derivative curves are superposed, reflecting the pressure response characteristics during well storage stage; (II) intermediate flow section (transient section between wellbore storage section and mid-radial flow section); (III) mid-radial flow section, where fluids flow of each layer achieves plane radial flow before crossflow occurs; (IV) crossflow section where fluids in low permeability layer transport through interlayer into high permeability layer; and (V) systematic radial flow section, where the whole system presents plane radial flow over time.

- The remarkable feature of the crossflow in type curves is the occurrence of "concave." The effect of polymer rheology on type curve section (V) is dramatically reduced by crossflow, which means the pressure curve and pressure derivative curve of polymer flooding are similar to those of water flooding in systematic radial flow section (V). Sensitivity analysis was performed to investigate the effect of different parameters on the type curves, including interporosity flow coefficient, formation coefficient ratio, storativity ratio, initial polymer concentration, IPV, and wellbore storage coefficient. The influence of IPV on the well testing in polymer flooding reservoirs can be neglected, since polymer flooding usually results in unremarkable IPV.

- Field tests were conducted in two wells of crossflow double-layer reservoirs by polymer flooding. The field test data were interpreted and history-matched by employing our well testing interpretation method, which indicated our model can accurately interpret field test data and evaluate formation. Moreover, formation damage caused by polymer flooding can also be evaluated by comparison of the interpreted permeability with initial layered permeability before polymer flooding. If interpreted permeability is much lower than initial permeability, specific techniques should be employed to eliminate formation damage and enhance oil recovery.

ACKNOWLEDGMENTS

The authors gratefully acknowledge financial support from the Science Foundation of China University of Petroleum, Beijing (no. YJRC-2013-41), the National Natural Science Foundation of China (no. 51304223), and the National Science and Technology Major Projects (no. 2011ZX05024-004-07).

REFERENCES

1.   B. Y. Jamaloei, R. Kharrat, and F. Torabi, "Analysis and correlations of viscous fingering in low-tension polymer flooding in heavy oil reservoirs," Energy and Fuels, vol. 24, no. 12, pp. 6384–6392, 2010.

2.   A. Bera, A. Mandal, and B. B. Guha, "Synergistic effect of surfactant and salt mixture on interfacial tension reduction between crude oil and water in enhanced oil recovery," Journal of Chemical & Engineering Data, vol. 59, no. 1, pp. 89–96, 2014.

3.   R. Farajzadeh, A. Ameri, M. J. Faber, D. W. Batenburg, D. W. Boersma, and J. Bruining, "Effect of continuous, trapped, and flowing gas on performance of Alkaline Surfactant Polymer (ASP) flooding,"Industrial & Engineering Chemistry Research, vol. 52, no. 38, pp. 13839–13848, 2012.

4.   H. Yu, C. Kotsmar, K. Y. Yoon et al., "Transport and retention of aqueous dispersions of paramagnetic nanoparticles in reservoir rocks," in Proceedings of the 17th SPE Improved Oil Recovery Symposium (IOR '10), pp. 1027–1047, Tulsa, Okla, USA, April 2010.

5.   T. Zhang, M. Murphy, H. Yu et al., "Investigation of nanoparticle

adsorption during transport in porous media," in Proceedings of the SPE Annual Technology Conference and Exibition, Society of Petroleum Engineers (SPE), New Orleans, La, USA, 2013, SPE paper no. 166346.

6.   B. S. Shiran and A. Skauge, "Enhanced oil recovery (EOR) by combined low salinity water/polymer flooding," Energy and Fuels, vol. 27, no. 3, pp. 1223–1235, 2013.

7.   G. Ren, H. Zhang, and Q. P. Nguyen, "Effect of surfactant partitioning between CO_2 and water on CO_2 mobility control in hydrocarbon reservoirs," SPE Paper 145102, Society of Petroleum Engineers (SPE), Kuala Lumpur, Malaysia, 2011.

8.   G. Ren, A. W. Sanders, and Q. P. Nguyen, "New method for the determination of surfactant solubility and partitioning between CO_2 and brine," The Journal of Supercritical Fluids, vol. 91, pp. 77–83, 2014.

9.   N. B. Wyatt, C. M. Gunther, and M. W. Liberatore, "Increasing viscosity in entangled polyelectrolyte solutions by the addition of salt," Polymer, vol. 52, no. 11, pp. 2437–2444, 2011.

10.   L. W. Lake, Enhanced Oil Recovery, Prentice Hall, Boston, Mass, USA, 1st edition, 1996.

11.   H. C. Lefkovits, P. Hazebroek, E. E. Allen, and C. S. Matthews, "A study of the behaviour of bounded reservoir composed of stratified layers," Society of Petroleum Engineers Journal, vol. 1, no. 1, pp. 43–58, 1961.

12.   S. M. Tariq and H. J. Ramey, "Drawdown behavior of a well with storage and skin effect communicating with layers of different Radii and other characteristics," SPE Paper 7453, Society of Petroleum Engineers (SPE), Houston, Tex, USA, 1978.

13.   F. Kucuk, M. Karakas, and L. Ayestaran, "Well testing and analysis techniques for layered reservoirs," SPE Formation Evaluation, vol. 1, no. 4, pp. 342–354, 1986.

14.   F. J. Kuchuk, P. C. Shah, L. Ayestaran, and B. Nicholson, "Application of multilayer testing and analysis: a field case," in Proceedings of the SPE Annual Technical Conference and Exhibition, SPE-15419-MS, Society of Petroleum Engineers (SPE), New Orleans, La, USA, October 1986.

15. D. Bourdet and F. Johnston, "pressure behavior of layered reservoirs with crossflow," SPE paper 13628, Society of Petroleum Engineers (SPE), Bakersfield, Calif, USA, 1985.

16. C. A. Ehlig-Economides and J. Joseph, "A new test for determination of individual layer properties in a multilayered reservoir," SPE Formation Evaluation, vol. 2, no. 3, pp. 261–282, 1987.

17. C.-T. Gao, "Single phase fluid flow in a stratified porous medium with crossflow," Society of Petroleum Engineers Journal, vol. 24, no. 1, pp. 97–106, 1984.

18. R. R. Jackson, R. Banerjee, and R. K. M. Thambynayagam, "An integrated approach to interval pressure transient test analysis using analytical and numerical methods," in Proceedings of the 13th Middle East Oil and Gas Show Conference, SPE Paper 81515, pp. 765–773, Society of Petroleum Engineers, Bahrain, June 2003.

19. M. M. Kamal, Y. Pan, J. L. Landa, and O. O. Thomas, "Numerical well testing: a method to use transient testing results in reservoir simulation," in Proceedings of the SPE Annual Technical Conference and Exhibition (ATCE '05), pp. 1905–1917, October 2005.

20. A. Mijinyawa, P. Alamina, and A. Orekoya, "An integrated approach to well test analysis-use of numerical simulation for complex reservoir systems," SPE paper 140617, Society of Petroleum Engineers (SPE), Warri, Nigeria, 2010.

21. L. Zhang, J. Guo, and Q. Liu, "A well test model for stress-sensitive and heterogeneous reservoirs with non-uniform thicknesses," Petroleum Science, vol. 7, no. 4, pp. 524–529, 2010.

22. S. K. Veerabhadrappa, J. J. Trivedi, and E. Kuru, "Visual confirmation of the elasticity dependence of unstable secondary polymer floods," Industrial and Engineering Chemistry Research, vol. 52, no. 18, pp. 6234–6241, 2013.

23. H. Zhang, R. S. Challa, B. Bai, X. Tang, and J. Wang, "Using screening test results to predict the effective viscosity of swollen superabsorbent polymer particles extrusion through an open fracture," Industrial and Engineering Chemistry Research, vol. 49, no. 23, pp. 12284–12293, 2010.

24. P. V. D. Hoek, H. Mahani, T. Sorop et al., "Application of injection fall-off analysis in polymer flooding," in Proceedings of

the SPE Europec/EAGE Annual Conference and Exhibition, SPE-154376-MS, Society of Petroleum Engineers (SPE), Copenhagen, Denmark, June 2012.

25. J. Jian, Q. Hou, L. Cheng, W. Liu, J. Li, and Y. Zhu, "Recent progress and effects analysis of surfactant-polymer flooding field tests in China," in Proceedings of the SPE Enhanced Oil Recovery Conference, SPE Paper 165213, Society of Petroleum Engineers (SPE), Kuala Lumpur, Malaysia, July 2013.

26. J. Liu, Y. Guo, J. Hu et al., "Displacement characters of combination flooding systems consisting of gemini-nonionic mixed surfactant and hydrophobically associating polyacrylamide for bohai offshoreoilfield," Energy and Fuels, vol. 26, no. 5, pp. 2858–2864, 2012.

27. J. Shi, A. Varavei, C. Huh, M. Delshad, K. Sepehrnoori, and X. Li, "Transport model implementation and simulation of microgel processes for conformance and mobility control purposes," Energy & Fuels, vol. 25, no. 11, pp. 5063–5075, 2011.

28. L. Ali and M. A. Barrufet, "Profile modification due to polymer adsorption in reservoir rocks," Energy & Fuels, vol. 8, no. 6, pp. 1217–1222, 1994.

29. N. Lai, X. Qin, Z. Ye, C. Li, K. Chen, and Y. Zhang, "The study on permeability reduction performance of a hyperbranched polymer in high permeability porous medium," Journal of Petroleum Science and Engineering, vol. 112, pp. 198–205, 2013.

30. A. L. Ogunberu and K. Asghari, Water Permeability Reduction under Flow-Induced Polymer Adsorption, SPE Paper 89855, Society of Petroleum Engineers (SPE), Houston, Tex, USA, 2004.

31. R. S. Seright, "Disproportionate permeability reduction with pore-filling gels," SPE Journal, vol. 14, no. 1, pp. 5–13, 2009.

32. P. L. Bondor, G. J. Hirasaki, and M. J. Tham, "Mathematical simulation of polymer flooding in complex reservoirs," Society of Petroleum Engineers Journal, vol. 12, no. 5, pp. 369–382, 1972.

33. P. J. Carreau, Rheological equations for molecular network theories [PhD dissertation], University of Wisconsin-Madison, Madison, Wis, USA, 1968.

34. D. M. Mete and R. B. Bird, "Tube flow of non-newtonian polymer solutions: part I. Laminar flow and rheological model," AIChE Journal, vol. 10, no. 6, pp. 878–881, 1964.

35. P. G. Flory, Principles of Polymer Chemistry, Cornell University Press, New York, NY, USA, 1953.

36. X. Wang, "Determination of the main parameters in the numerical simulation of polymer flooding,"Petrol Exploration Development, vol. 3, pp. 69–76, 1990.

37. J. Wang, Physic-Chemical Fluid Mechanics and Application in Chemical EOR, Petroleum Industry Press, Beijing, China, 2008.

38. D. Bourdet, J. A. Ayoub, and Y. M. Pirard, "Use of pressure derivative in well test interpretation," SPE Journal, vol. 4, no. 2, pp. 293–302, 1989.

Chapter 2

Signal Feature Extraction and Quantitative Evaluation of Metal Magnetic Memory Testing for Oil Well Casing Based on Data Preprocessing Technique

Zhilin Liu[1], Lutao Liu[2], and Jun Zhang[3]

[1]College of Automation, Harbin Engineering University, Harbin, Heilongjiang 150001, China

[2]College of Information and Telecommunication, Harbin Engineering University, Harbin, Heilongjiang 150001, China

[3]School of Electrical and Information Engineering, Jiangsu University, Zhenjiang, Jiangsu 212013, China

ABSTRACT

Metal magnetic memory (MMM) technique is an effective method to achieve the detection of stress concentration (SC) zone for oil well casing. It can provide an early diagnosis of microdamages for preventive protection. MMM is a natural space domain signal which is weak and vulnerable to noise interference. So, it is difficult to achieve effective feature extraction of MMM signal especially under the hostile subsurface environment of high temperature, high pressure, high humidity, and multiple interfering sources. In this paper, a method of median filter preprocessing based on data preprocessing technique is proposed to eliminate the outliers point of MMM. And, based on wavelet transform (WT), the adaptive wavelet denoising method and data smoothing arithmetic are applied in testing the system of MMM. By using data preprocessing technique, the data are reserved and the noises of the signal are reduced. Therefore, the correct localization of SC zone can be achieved. In the meantime, characteristic parameters in new diagnostic approach are put forward to ensure the reliable determination of casing danger level through least squares support vector machine (LS-SVM) and nonlinear quantitative mapping relationship. The effectiveness and feasibility of this method are verified through experiments.

INTRODUCTION

Caused by the factors of erosion, geology, and engineering, well casing damage leads to huge economic losses in oil field every year because of the long-term nonuniform load on downhole casing which results in severe local stress concentration and bending, deformation, and breaking [1–3]. Therefore, it is one of the difficulties in nondestructive testing to predict the abnormal stress concentration of oil well casing in order to prevent casing damage. Stress is regarded as one of the major factors affecting ferromagnetic behavior, along with magnetic field and temperature. Effect of stress or strain on magnetization is called the Villari effect, inverse magnetostrictive effect, or piezomagnetism. In general, it is simply referred to as magnetomechanical effect. Since an applied stress can alter the domain structure and have a substantial effect on the low-field magnetic properties, such as remanence and permeability, recently these effects are mostly found in practical applications of magnetic nondestructive testing, actuators, and magnetic sensors. As a result, the effects have been paid considerable attention in the literature [1–4]. Nevertheless, the coupling effect between mechanical and magnetic properties is so complicated to stunt the development of these properties in nondestructive testing application.

Traditional nondestructive testing such as radiographic testing, ultrasonic testing, penetrant testing, magnetic particle testing, and eddy current testing is effective in detecting existing cracks, which means that the detection of crack in the bud can not be achieved in time before crack develops severely. However, MMM technique is a new method of nondestructive testing which can accomplish early diagnosis and detection of defects. With the testing characteristic of stress concentration, it is capable of detecting the most dangerous damage in advance. It has been widely applied in the field of electric power, railway, and petrochemical industry and has been proved effective. Despite of all these, MMM is found to be a natural space domain signal and have the same order of magnitude as the earth magnetic field, which means it is of random nonsmooth signal as well as low signal-to-noise ratio and is vulnerable to noise interference [4]. Meanwhile, quantitative evaluation is affected because it is difficult to correctly extract the features of MMM signal under the hostile subsurface environment, such as high temperature, high pressure, and great noise. Another difficulty for feature extraction is that underground casing is wrapped in thermal-protective coating of a large damping, which results in producing weaker signal or even no signal from the stress concentration zone where cracks appear.

Data preprocessing and data driven is the first step in diagnosis for casting by using MMM. The reliability, usability, and integrality of measurement data can affect the accuracy, efficiency, and effectiveness of model reconstruction directly. The concept of data preprocessing and data driven is often used in computer science. But because of the exceeding progress of computer techniques, massive process data can be obtained by the intelligentized industry, so data preprocessing and data driven have been taking up a lot of attentions in engineering science. Rapid improvement of database capacity make people use data more effectively and fulfill more functions. Data means information, so the so-called data preprocessing and data driven are drawing information from data and use information to realize different objects.

To draw information from data, statistical techniques are the chief method, and applications based on multivariate statistical techniques become the main part of data preprocessing and data-driven area. At present, there have been many papers about the application in different fields of industry based on data-driven algorithm and techniques [5–8]. Based on the basic data-driven methods for process monitoring and fault diagnosis, a comparison study in [9] is provided. Authors in [9] illustrated the efficiencies of data-driven methods discussed in their paper through the application of an industrial benchmark of Tennessee Eastman (TE) process.

The most important mathematical tools in statistical techniques, filtering algorithm, and wavelet analysis are commonly used in data preprocessing and data-driven field. In the process of data acquisition, redundancy, and measurement, noise will be introduced inevitably. These error points can bring great impacts on the model reconstruction and analysis of data feature. In order to extract the feature of data better, data filtering must be applied to make the errors removed. Authors in [10] proposed an intelligent data filtering method based on artificial neural networks to detect bearing defects of induction motors. In [11], a robust H_∞ filtering problem is investigated for a class of complex network systems which has stochastic packet dropouts and time delays, combined with disturbance inputs. Authors in [12] deal with the design problem of minimum entropy H_∞ filter in terms of linear matrix inequality (LMI) approach for linear continuous-time systems with a state-space model subject to parameter uncertainty that belongs to a given convex bounded polyhedral domain. In [13], a filtering algorithm for maneuvering target tracking is presented based on smoothing spline fitting.

The wavelet analysis theory is gradually developing to become one of important technologies in the dynamic measurement signal process field by its advantage of multiresolution and multidimensional analysis on time frequency. Authors in [14] propose control strategy for the energy management which is based on the combination of wavelet transform and neural network arithmetic. In the control strategy proposed, wavelet is in charge of decomposing and then reconfiguring the power difference between generated power and consumed power by loads. In [15], authors address an application of wavelet networks in identification and control design for a class of structures equipped with a type of semiactive actuators. The wavelet analysis theories and methods are developing and are far from maturation. Wavelet analysis and its application have great potentialities in many applied fields of natural science, and its application in MMM test and estimating stress concentration zone is increasing.

In this paper, according to the foregoing reference, a method of multisource information processing and multifeature quantitative evaluation is described to solve the problems of MMM signal processing of underground casing. First of all, median filter preprocessing is adopted to eliminate the outliers point of MMM. Secondly, based on wavelet transform (WT), the estimation is made by adaptive threshold of wavelet transform coefficients with various scaled space signals through optimal soft threshold denoising, which restrains the noise signal to extract gradient and zero-crossing point features to achieve correct localization of SC zone. Thirdly, new diagnostic approach of combined characteristic parameters is put forward to ensure the reliable determination of casing danger level through LS-SVM and nonlinear quantitative mapping relationship. The effectiveness and feasibility of this method are verified through experiments.

MECHANISM OF METAL MAGNETIC MEMORY TESTING

The studies of modern material science and ferromagnetic have proved that if iron artifacts are influenced by its working load under geomagnetic environment, their interior structure will show magnetostrictive magnetic domain orientation and irreversible reorientation. Theoretically, relationship between the magnetic field leakage of (H_p) and the changes of mechanical stress ($\Delta\sigma$) of ferromagnetic artifacts under test is as follows [2, 3]:

$$H_p = \frac{\lambda^H}{\mu_0}\Delta\sigma, \tag{1}$$

where λ^H is an irreversible component in magnetoelastic effect, and it is a function that depends on mechanical stress as well as the intensity and temperature of the external magnetic field; $\mu_0 = 4\pi \times 10^{-7}$ is the permeability of vacuum. Metal magnetic memory theory has proved that the maximum variation of scattered magnetic leakage field H_p occurs in stress concentration and deformation zone; that is, the tangential component of magnetic leakage $H_p(x)$ shows the maximum value while normal component of the magnetic leakage $H_p(y)$ is shown to be zero (Figure 1). This irreversibility of magnetic state will remain after the elimination of working load, which makes it possible to accomplish the accurate diagnosis of component defects and (or) stress concentration zone through the determination of normal component $H_p(y)$ in the magnetic leakage of scattered magnetic leakage field. In [16, 17], the absolute value of maximum gradient value in magnetic variation is taken as a diagnostic parameter to estimate stress concentration level (a patent belongs to Energodiagnostika Co. Ltd., of Russia), that is, $K = \left| dH_p(y) / dx \right|$, is taken as a measurement indicator. The features of zero-crossing point and gradient value are two key characteristic parameters in MMM testing technique. Experts from Energodiagnostika Co., Ltd., have proposed supplementary rules at the MMM application conference in Anshan in 2004, which is mainly about the comprehensive localization of stress concentration zone through maximum gradient area and zero-crossing point area [1–4]. Though it is easy to locate stress concentration zone according to the smooth MMM curves like the ones in Figure 1, the actual curves collected are obviously much more complicated than what is in Figure 1, because noise variation is often included when differential derivative technique is employed, which makes it hard to extract the accurate gradient value of signal saltation. In addition, subsurface environment is extremely hostile and has the characteristics of high temperature, high pressure, high humidity, and great noise; underground casing is wrapped in thermal-protective coating of a large damping, which results in worse signal-to-noise ratio of weak MMM signal or even totally overwhelmed by noise. Therefore, it is difficult to extract gradient characteristic value of MMM as well as achieve quantitative evaluation of danger level, which proves the necessity of MMM signal processing.

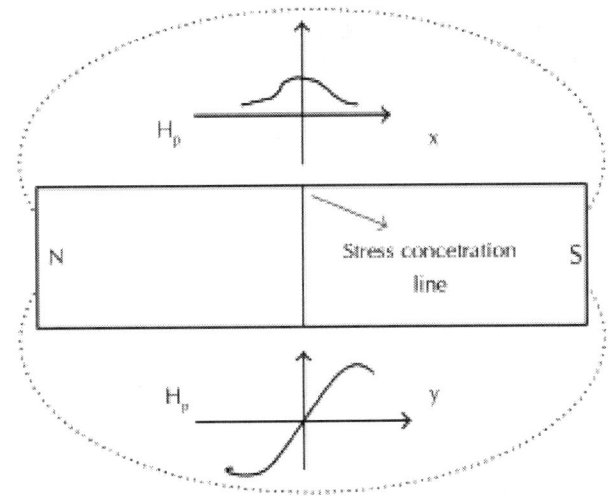

Figure 1: Principle diagram of MMM testing.

SIGNAL ANALYSIS OF MMM TESTING FOR OIL WELL CASING

While variable characteristics of MMM signal in tensile tests have been introduced in most of the present research papers [18, 19], few extrusion tests have been involved. The main function of oil well casing is pressure-bearing. In this text, a ground test was carried out in the simulation well of Daqing oilfield. During the test, short oil well casings with the length of 1 m were truncated, respectively, from a 11-meter long oil well casing (specimen is about 14 mm; wall thickness is 7.6 mm) and were labeled as specimen 1 and specimen 2. The two specimens were placed, respectively, in the middle of hydraulic platform of a NYL-300 compression testing machine, and stress tests were carried out in the middle part of the specimens in order to eliminate the effect of end face. The nominal working strengths were 20, 40,…. and 240 kN. Stress application was done every other 20 kN and lasted for 10 minutes each time. After each stress application, the specimen was removed and was vertically placed at a fixed position. The experimental facility is shown in Figure 2.

Figure 2: Stress test facility.

Before stress tests, there was no stress concentration zone in oil well casing; see the MMM curve in Figure 3. There is no obvious peak-peak singular feature of local gradient and signal. When the working strength is equal to or larger than 40 kNs pressure, the curve showed significant change, that is, obvious peak-peak value in Figure 4 and maximum gradient area after Fourier analysis of the curve. See Figure 5. This proves that MMM signal energy is mainly concentrated in low-frequency area. From Figure 6, it can be seen that in several areas, the gradient values are above 8 or close to 8. According to the criterion of Russian patent, this means there is dangerous stress concentration zone (oil well casing is close to hazardous situation when gradient values are above 8). However, after the experiment, conclusion has been drawn that maximum stress concentration level only appears intensively in the intermediate region, which means there should be low stress concentration level in other areas and the utilization will not be affected. Therefore, it is difficult to determine danger level depending only on differential derivative [20]. It is necessary to conduct MMM signal processing, especially under more complicated underground environment.

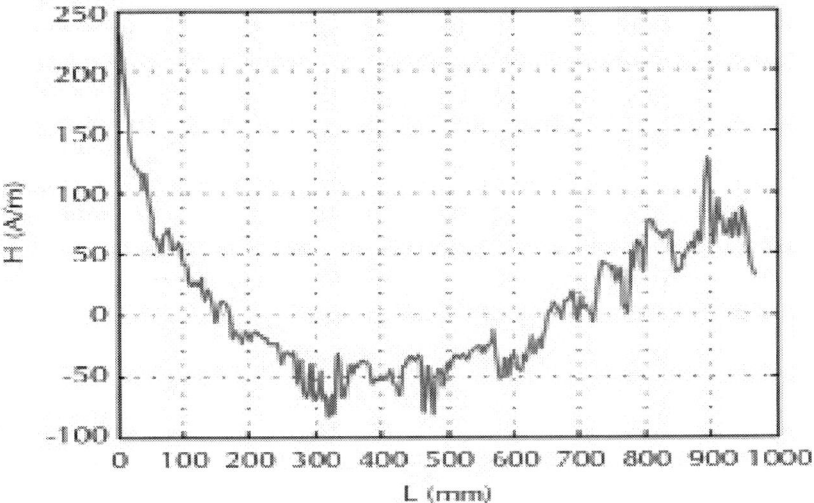

Figure 3: MMM signal of casing before stress test.

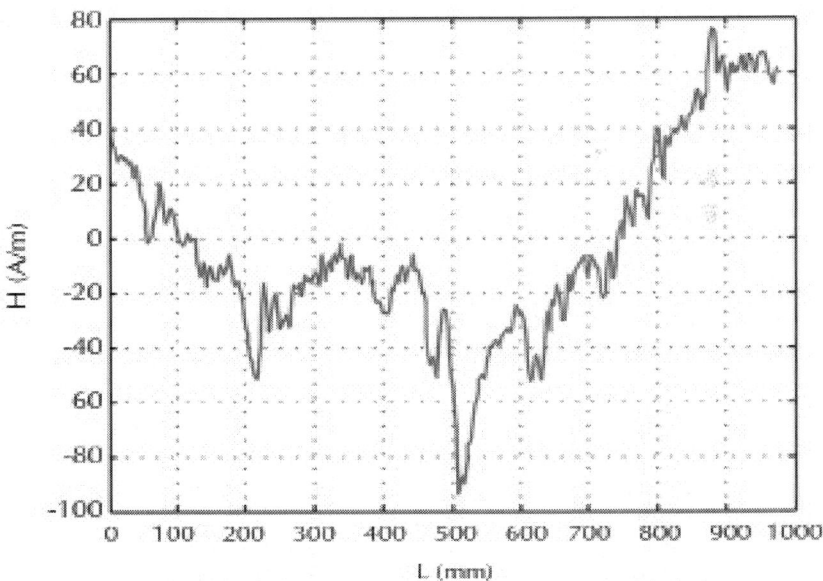

Figure 4: MMM signal of casing after stress test.

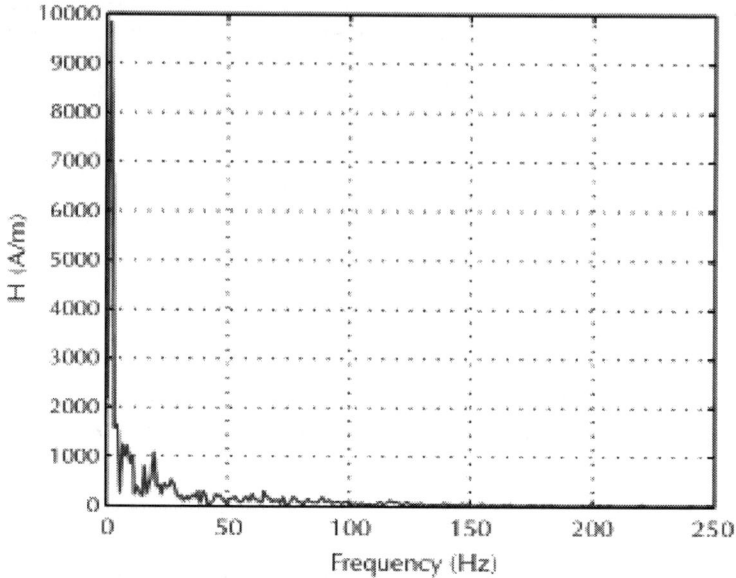

Figure 5: MMM signal frequency.

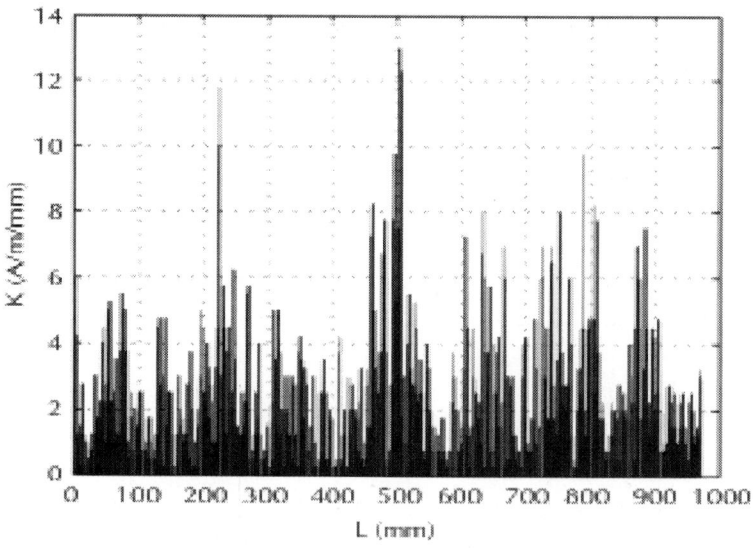

Figure 6: Gradient value variation of MMM signal.

MMM SIGNAL PROCESSING AND FEATURE EXTRACTION FOR OIL WELL CASING

There is a great influence of underground interference and noise on MMM data, and the main source of noise is found to be the high-frequency noise of measurement noise and the interfering signal from probe vibration. Since MMM signal is random signal, strictly speaking, it lacks stability. Though Fourier analysis can achieve a general analysis on the spectrum features of MMM signal, it does not possess local analysis feature of time domain and frequency domain [20]. Therefore, wavelet analysis is adopted in this section to process MMM signal.

Data Smoothing

Since MMM signal is weak spatial domain signal with low frequency [21, 22], it is necessary to firstly accomplish the smoothing of data collected in order to remove possible interference signal and meaningless isolated outliers point. To ensure high fidelity of signal amplitude as well as real time of the testing system without producing new quantization parameters, median smooth filter is adopted. Consider

$$y(m) = \text{Median}\left[x(m), x(m-1), x(m-2)\right],$$

(2)

Where y (m) is the output, x (m) is the input signal sequence in spatial domain, and Median is median function.

Wavelet Analysis

Since wavelet transformation (MT) has a good multiresolution time-frequency analysis feature which is capable of reducing noise as well as retaining the edge, it has become the key method of the extraction of MMM signal singularity [23–26]. For MMM signal f (t) which contains noise, the model in wavelet domain is

$$y(t) = f(t) + \delta \cdot n(t),$$

(3)

Where n(t) represents Gaussian white noise and σ_n indicates noise intensity. $E[n(t)]$ Shows the mathematical expectation of random variable; thus,

$$E[n(t)] = 0, \qquad E[n(u)n(v)] = \begin{cases} \sigma_n^2, & u = v, \\ 0, & u \neq v. \end{cases}$$

(4)

Define $W_s(n(t))$ as the wavelet transform value of n(t). To some degree, it is also a random variable of t. Under wavelet decomposition scale s, there is

$$\left| W_s(n(t)) \right|^2 = \iint_{-\infty}^{+\infty} n(u)n(v)\,\psi(X-u)\,\psi(X-v)\,du\,dv.$$

(5)

Its mathematical expectation is

$$E\left(\left| W_s(n(t)) \right|^2\right)$$

$$= \iint_{-\infty}^{+\infty} \sigma_n^2 \delta(u-v)\,\psi(X-u)\,\psi(X-v)\,du\,dv$$

$$= \frac{\|\psi\|^2 \sigma_n^2}{s}.$$

(6)

This indicates that the average density of white noise modulus maximum is inverse proportion to the scale; that is, the greater the scale is, the sparser the modulus maximum will be. While WT of noise on different scales is highly irrelevant, WT of MMM singular signal often has a strong correlation; that is, local modulus maximums on adjacent scales almost share the same position and the same sign, which is the theoretical foundation of MMM signal processing.

Wavelet-Based Adaptive Threshold Denoising

The principle of wavelet-based denoising is to accomplish a wavelet analysis on measurement signal mixed with noise and to separate them according to the different characteristics between signal and noise under WT. The wavelet coefficients which belong to noise are set to zero; wavelet reconstruction is carried out on the left to get useful signal. Wavelet coefficients obtained from hard threshold method are discrete [22], which often result in oscillation effect for signal reconstruction. Soft threshold function can achieve smooth denoising of wavelet coefficients in low scales. However, in high scales, it will cause a decline of signal-to-noise ratio. Therefore, it is necessary to contract wavelet coefficients in low scales as well as protect wavelet coefficients in high scales in order to restrain the oscillation.

The combination of soft and hard threshold methods is adopted, that means that, under the premise of maximum denoising [23], errors are reduced to the greatest extent with the purpose of achieving optimal denoising. Consider

$$\begin{cases} \operatorname{sgn}\left(W_{s,k}\right)\left(\left|W_{s,k}\right| - a \cdot \sigma\right), & \left|W_{s,k}\right| \geq \sigma, \\ 0, & \left|W_{s,k}\right| < \sigma, \end{cases}$$

$$0 \leq a \leq 1. \tag{7}$$

When $a=0$, it is hard threshold method; when $a=1$, it is soft threshold method; $W_{s,k}$ represents wavelet coefficient. To get optimal estimation of the signal, the value of a is determined by minimum mean square error criterion. Optimal threshold value a are [23]

$$a = E\left[|n|\right] \cdot \frac{P_+ - P_-}{P_+ + P_-},$$

$$P_+ = \Pr\left\{\mathrm{sgn}(y) = \mathrm{sgn}(n), |y| > \sigma\right\},$$

$$P_- = \Pr\left\{\mathrm{sgn}(y) \neq \mathrm{sgn}(n), |y| > \sigma\right\}. \tag{8}$$

For MMM signal which has undergone filtering preprocessing, the noise is mainly Gaussian white noise; thus, $E\left[|n|\right] \approx 0.6744$ and a=0.6744.

Under regular denoising method, universal threshold value is $\sigma = \sigma_n \sqrt{2 \log N}$; N represents signal length. According to

$$\sigma_s = 2^{1-(\sigma_s/\sigma_n) \cdot s} \cdot \sigma, \tag{9}$$

it can be seen that actually noise threshold will decline along with the increase of scales. So, it is not inadvisable to choose the fixed threshold value. If the threshold value is too large, useful signal is prone to be filtered out, which will have an impact on signal-to-noise ratio. So, adaptive threshold value is chosen by (9), which changes according to the scales and standard deviation. Noise standard deviation is estimated as follows:

$$\sigma_n = \frac{\mathrm{Median}\left(|w_{1,j}|\right)}{0.6745}. \tag{10}$$

Since WT is linear transformation, both signal and noise wavelet coefficients accord Gaussian distribution, and signal standard deviation of each wavelet scale is estimated in terms of approximate maximum probability:

$$\sigma_s^2 = \max\left(0, \frac{1}{n^2}\sum_{j=1}^{n} w_{s,t}^2 - \sigma_n^2\right).$$

(11)

Therefore, new wavelet coefficients are obtained, and signal reconstruction is achieved through inverse transforms.

Signal denoising procedures are as follows:

- wavelet coefficients are obtained through six-layer decomposition according to Db10 wavelet mother function;
- based on (9), the first four-layer threshold values are determined, and optimal threshold value (7) is employed to process wavelet coefficients to get new wavelet coefficients;
- signal is reconstructed through inverse transforms.

Data from Figure 4 is applied to adaptive threshold denoising; see Figures 7 and 8. It is obvious that gradient value maximum of oil well casing in the middle area is 8, and the stress concentration level is the highest, which is in accordance with experimental results and should be well focused on. Gradient values in other areas are less than 4; the stress concentration zone is relatively small. Therefore, gradient values after signal denoising are more accessible to qualitative and quantitative evaluation on service life of oil well casing for influence of noise is eliminated.

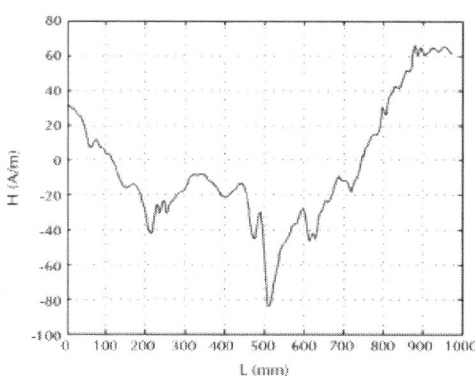

Figure 7: MMM signal after adaptive denoising.

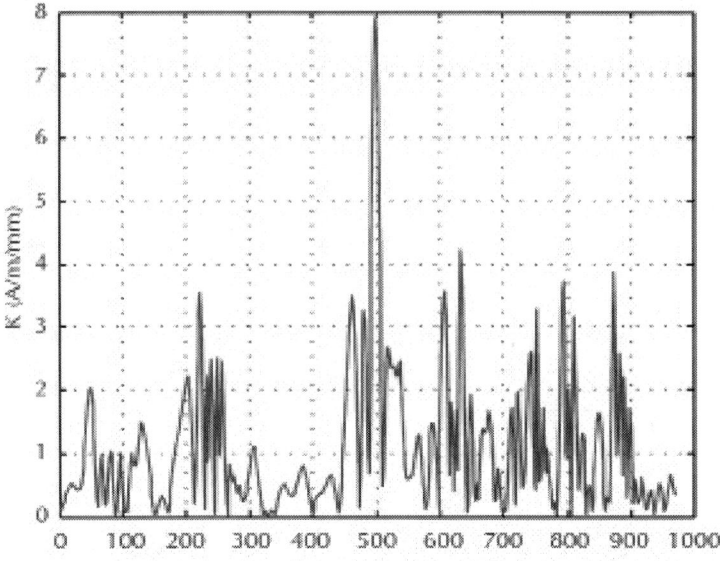

Figure 8: MMM gradient value after adaptive denoising.

After the denoising reconstruction of MMM signal collected, feature extraction should be accomplished. Patented technology raised by Energodiagnostika Co., Ltd., of Russia is to determine stress concentration zone through gradient value maximum and zero-crossing point. Gradient value is considered as characteristic parameter, which has been pointed out in existing researches that one single parameter often causes misjudgment in real. In this section, peak-peak value and gradient value are chosen to be characteristic parameters as a whole with the purpose of ensuring accurate determination of stress concentration zone.

Signal peak-peak value PP0 is as follows: peak-peak value is taken as characteristic parameter, which eliminates the impact of signal baseline and enhances the reliability of detection, because it is an important distribution feature of signal detection. When peak-peak value is calculated, maximum and minimum values of signal are firstly sought, and then absolute value (range) of the difference between these two adjacent extremums is obtained, which is easy to achieve through computer. Consider

$$PP0 = \max\left[f\,(m)\right] - \min\left[f\,(m)\right],$$

(12)

where $\max[f(m)]$ and $\min[f(m)]$ represent a pair of adjacent extremums. During the experiment, peak-peak value is the key index.

$Z = [T,PP0]^{\mathsf{T}}$ Refers to the feature vector; T is gradient value. When both of the two characteristic component values are larger than the corresponding threshold value, there is the real dangerous area of stress concentration zone. If gradient value is larger than threshold value while peak-peak value is relatively small, there is not necessarily a stress concentration zone but possibly an uncertain area created by noise. Accurate determination needs to be done with the help of other nondestructive testing methods. Qualitative analysis is only a way to determine stress concentration zone of oil well casing; it is unable to satisfy the need of the quantitative evaluation on danger level, which will be in the next section.

QUANTITATIVE EVALUATION ON DANGER LEVEL OF OIL WELL CASING

Mechanism model which reflects MMM effect is the quantitative basis of nondestructive testing. Present mechanism studies are mainly based on theories such as common effect of stress and external magnetic field, stress magnetization of energy maximum principle, and stress permeability effect. Combined with test data, MMM mechanism model is established from various angles; however, complete and rigorous theoretical system has not been formed yet. So, the scope of applications is limited. In the meantime, local elastic and plastic deformation of ferromagnetic material has been noticed during the experiment, and nonlinear variation of normal magnetic flux leakage happened on the surface of local variation area. Therefore, feature parameter and stress concentration of MMM signal are of unknown mathematical relation, which requires a solution of quantitative evaluation problem through nonlinear modeling technology. Gradient value and peak-peak value as combined vector are taken to describe signal features and there are two input ends. Output ends of quantitative model are also two. According to the safe requirement of real engineering oil well casing, "00" represents the fact that elastic deformation is relatively small, which has no influence on using; "01" shows a relatively large elastic deformation, which requires regular inspection; "10" means that critical plastic changing area, protective measures should be carried out; "11" indicates that severe deformation of oil well casing has happened and they must be replaced.

Since $\{T,PP0 \rightarrow Y\}$ obtained from experiment is small sample, quantitative classification of danger level through small sample study method is adopted in this section, and nonlinear mapping relationship is established through LS-SVM. LS-SVM is a machine learning algorithm based on statistical learning theory. Learning according to the principle of structural risk minimization, it translates optimization problem into a problem of convex quadratic programming, which ensure that the extreme solution is the globally optimal solution [27].

A training sample set is taken into consideration which includes N data point $\{x_k, y_k\}, k=1,2,....,N, x_k \in R^n$ which represents the input sample and $y_k \in R$ shows the output sample. Regression model is as follows:

$$y(x) = w^T \varphi(x) + b.$$

(13)

In the above formula, nonlinear mapping $\varphi(x)$ map input data to high-dimension feature space, which makes the nonlinear regression problem in the original space to translate into linear regression problem in feature space. And $w \in R^n$, $b \in R$.

The definition of objective function in regression LS-SVM is

$$\min_{w,e} J(w,e) = \frac{1}{2}w^T w + \frac{C}{2}\sum_{k=1}^{N} e_k^2,$$

$$y_k = w^T \varphi(x_k) + b + e_k, \quad k = 1, 2, \ldots, N,$$

(14)

k=1,2,...,N, C represents the constant which is called penalty factor and it is used to balance model complexity and fitting precision, and e_k refers to the error term. Constrained optimization problem is translated into unconstrained optimization problem, and Lagrange multiplier $_k$ is introduced, the corresponding Lagrange function is

$$L\left(w, b, e, \alpha\right)$$

$$= J\left(w, e\right) - \sum_{k=1}^{N} \alpha_k \left\{ w^T \varphi\left(x_k\right) + b + e_k - y_k \right\}.$$ (15)

Under Karush-Kuhn-Tucker condition [16], we can get

$$\frac{\partial}{\partial w} L = 0 \longrightarrow w = \sum_{i=1}^{N} \alpha_k \varphi\left(x_k\right),$$

$$\frac{\partial}{\partial b} L = 0 \longrightarrow \sum_{i=1}^{N} \alpha_k = 0,$$

$$\frac{\partial}{\partial e_k} L = 0 \longrightarrow \alpha_k = Ce_k,$$ (16)

$$\frac{\partial}{\partial \alpha_k} L = 0 \longrightarrow w^T \varphi\left(x_k\right) + b + e_k - y_k = 0,$$

$$\begin{bmatrix} 0 & \tilde{1}^T \\ \tilde{1} & \Omega + C^{-1}I \end{bmatrix} \begin{bmatrix} b \\ \alpha \end{bmatrix} = \begin{bmatrix} 0 \\ Y \end{bmatrix},$$ (17)

Where $Y = [y_1, y_2, ..., y_N]^T$, $\tilde{1} = [1, 1, ..., 1]^T$, $\alpha = [\alpha_1, \alpha_2, ..., \alpha_N]^T$, and $\Omega(k,l) = \varphi(x_k)^T \varphi(x_1) = K(x_k, x_1), k, l = 1, 2, ..., N$. According to formula of (15) and (16), the regression model in which LS-SVM is applied to parameter estimation is

$$y = \sum_{i=1}^{N} \alpha_k K(x, x_k) + b.$$

(18)

In the above formula, α_k, b are the solution to formula (16).

Table 1 is the results of danger level classification for oil well casing, in which the first ten groups are training samples and the last two are testing samples. Therefore, quantitative evaluation of danger level for oil well casing is achieved through support vector machine.

Table 1: Quantitative evaluation on danger level for oil well casing

Serial number	Gradient value (A/m/mm)	Peak-peak value (A/m)	Output classification
1	4	130	00
2	5	150	00
3	8	210	01
4	9	260	01
5	9	230	01
6	14	300	10
7	10	230	10
8	12	250	10
9	18	280	11
10	14	270	11
11	7	200	01
12	10	286	10

Notation. Accurate classification requires a large quantity of training samples. Samples given here are related to the model number of MMM device (has an influence on the sensitivity of characteristic parameter) and material and model number of oil well casing. Experimental facility of underground test in Daqing Oilfield (located in Heilongjiang Province in China) is shown in Figures 9 and 10, and research results are the same as ground ones; no repeat will be done here. The casing after MMM testing is shown in Figure 11. As the casing cube is nondestructive, the MMM testing technique is suitable for practical application.

Figure 9: Probe device of underground MMM testing.

Figure 10: Sketch of scan test.

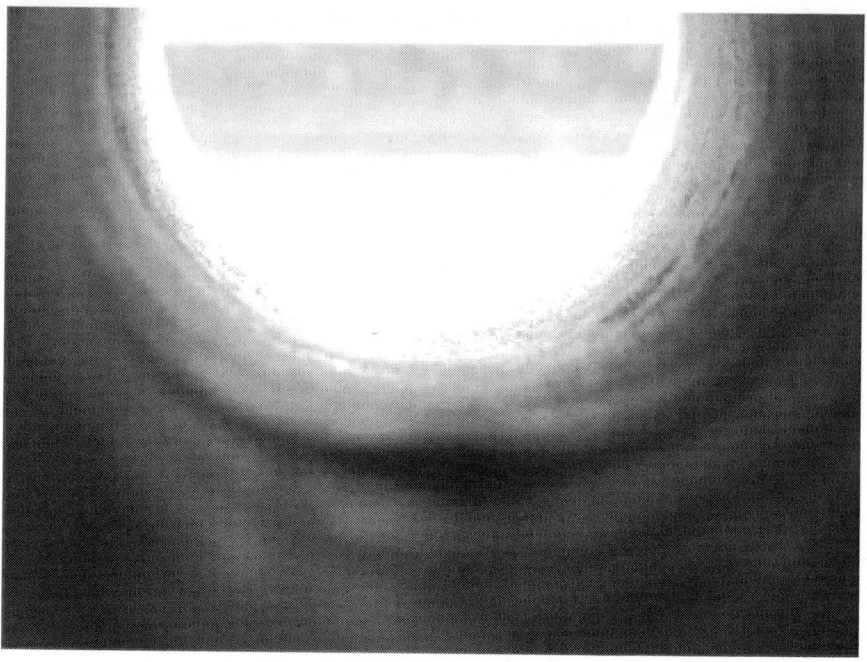

Figure 11: The casing after MMM testing.

CONCLUSIONS

The experiment indicates that MMM is capable of achieving an effective prediction on underground stress concentration zone for oil well casing. Due to the complicated underground environment and various interferences, digital processing technology is introduced to MMM analysis software in order to enhance signal-to-noise ratio as well as eliminate high-frequency noise. In addition, smooth data from filter processing achieve accurate feature extraction of MMM signal. At the same time, new combined feature vector is put forward, which is used to closely approach the nonlinear relationship of magnetic flux leakage signal and danger level through SVM. Therefore, the stress concentration level for oil well casing could be predicted in a timely and reliable manner. And it should also be noted that MMM is influenced by temperature greatly. With the oil well casing drilling into the ground, geothermal temperature will inevitably have an impact on the MMM signals. In the future work, efforts will be made to design the robust hardware to reduce the temperature drift. It is also a challenge to use advanced data-driven technology to deal with the complex MMM signals and remove the disturbance effect of geothermal temperature.

ACKNOWLEDGMENTS

This work is partially supported by National Natural Science Foundation of China (51379044), Fundamental Research Funds for the Central Universities (HEUCFX41304), and Heilongjiang Province Natural Science Foundation Projects (F200916).

REFERENCES

1.   A. A. Doubov, "Diagnostics of equipment and constructions strength with usage of magnetic memory,"Inspection Diagnostics, no. 6, pp. 19–29, 2001.

2.   A. A. Dubov and K. Sergey, "The metal magnetic memory method application for online monitoring of damage development in steel pipes and welded joints specimens," Welding in the World, vol. 57, no. 1, pp. 123–136, 2013.

3.   A. A. Doubov, "Express method of quality control of a spot resistance welding with usage of metal magnetic memory," Welding in the World, vol. 46, no. 6, pp. 317–320, 2002.

4.   A. A. Dubov, "Development of a metal magnetic memory method," Chemical and Petroleum Engineering, vol. 47, no. 11, pp. 837–839, 2012.

5.   S. Yin, G. Wang, and H. R. Karimi, "Data-driven design of robust fault detection system for wind turbines," Mechatronics, vol. 24, no. 4, pp. 298–306, 2014.

6.   S. Yin, S. X. Ding, A. H. A. Sari, and H. Hao, "Data-driven monitoring for stochastic systems and its application on batch process," International Journal of Systems Science, vol. 44, no. 7, pp. 1366–1376, 2013.

7.   S. Yin, S. X. Ding, X. Xie, and H. Luo, "A review on basic data-driven approaches for industrial process monitoring," IEEE

Transactions on Industrial Electronics, vol. 61, no. 11, pp. 6418–6428, 2014.

8.   S. Yin, H. Luo, and S. X. Ding, "Real-time implementation of fault-tolerant control systems with performance optimization," IEEE Transactions on Industrial Electronics, vol. 61, no. 5, pp. 2402–2411, 2014.

9.   S. Yin, S. X. Ding, A. Haghani, H. Hao, and P. Zhang, "A comparison study of basic data-driven fault diagnosis and process monitoring methods on the benchmark Tennessee Eastman process," Journal of Process Control, vol. 22, no. 9, pp. 1567–1581, 2012.

10.   J. Zarei, M. A. Tajeddini, and H. R. Karimi, "Vibration analysis for bearing fault detection and classification using an intelligent filter," Mechatronics, vol. 24, no. 2, pp. 151–157, 2014.

11.   J. Zhang, M. Lyu, H. R. Karimi, P. Guo, and Y. Bo, "Robust $H\infty$ filtering for a class of complex networks with stochastic packet dropouts and time delays," The Scientific World Journal, vol. 2014, Article ID 560234, 11 pages, 2014.

12.   J. Zhang, H. R. Karimi, Z. Zheng, M. Lyu, and Y. Bo, "$H\infty$ filter design with minimum entropy for continuous-time linear systems," Mathematical Problems in Engineering, vol. 2013, Article ID 579137, 9 pages, 2013.

13.   Y. Liu, J. Suo, H. R. Karimi, and X. Liu, "A filtering algorithm for maneuvering target tracking based on smoothing spline fitting," Abstract and Applied Analysis, vol. 2014, Article ID 127643, 6 pages, 2014.

14.   Y. D. Song, Q. Cao, X. Du, and H. R. Karimi, "Control strategy based on wavelet transform and neural network for hybrid power system," Journal of Applied Mathematics, vol. 2013, Article ID 375840, 8 pages, 2013.

15.   H. R. Karimi, M. Zapateiro, and N. Luo, "Application of adaptive wavelet networks for vibration control of base isolated structures," International Journal of Wavelets, Multiresolution and Information Processing, vol. 8, no. 5, pp. 773–791, 2010.

16.   M. Roskosz, "Metal magnetic memory testing of welded joints of ferritic and austenitic steels," NDT & E International, vol. 44, no. 3, pp. 305–310, 2011.

17.   T. Yan, J. Zhang, G. Feng, and J. Chen, "Inspection of wet steam

generator tubes based on metal magnetic memory method," Procedia Engineering, vol. 15, pp. 1140–1144, 2011.

18. Z. D. Wang, K. Yao, B. Deng, and K. Q. Ding, "Theoretical studies of metal magnetic memory technique on magnetic flux leakage signals," NDT & E International, vol. 43, no. 4, pp. 354–359, 2010.

19. Z. D. Wang, K. Yao, B. Deng, and K. Q. Ding, "Quantitative study of metal magnetic memory signal versus local stress concentration," NDT & E International, vol. 43, no. 6, pp. 513–518, 2010.

20. K. Yao, B. Deng, and Z. D. Wang, "Numerical studies to signal characteristics with the metal magnetic memory-effect in plastically deformed samples," NDT & E International, vol. 47, pp. 7–17, 2012.

21. J. Leng, Y. Liu, G. Zhou, and Y. Gao, "Metal magnetic memory signal response to plastic deformation of low carbon steel," NDT & E International, vol. 55, pp. 42–46, 2013.

22. X. Hai-Yan, W. Wen-Jiang, W. Ri-Xing, X. Min-Qiang, and L. Xue-Feng, "Stress distribution testing of 50 MW turbine fracture blade with metal magnetic memory method," Proceedings of the Chinese Society of Electrical Engineering, vol. 26, no. 4, pp. 72–76, 2006.

23. H. Y. Kang, Y. Lu, and W. Y. Shuzi, "Some algorithms for nondestructive testing of wire ropesłsignal pre-processing and character extraction," Nonde Structive Testing, vol. 22, no. 11, pp. 483–488, 2000.

24. J. Zhang, B. Wang, and B. Ji, "Signal processing for metal magnetic memory testing of borehole casing based on wavelet transform," Acta Petrol Ei Sinica, vol. 27, no. 2, pp. 137–140, 2006.

25. Z. Xiaoyong and Y. Yinzhong, "Multi-fault diagnosis method on Mallat pyramidal algorithm wavelet analysis," Control and Decision, vol. 19, no. 5, pp. 592–594, 2004.

26. R. Manojit, V. Kumar, B. D. Kulkarni, J. Sanderson, M. Rhodes, and M. Vander Stappen, "Simple denoising algorithm using wavelet transform," American Institute of Chemical Engineers Journal, vol. 45, no. 11, pp. 2461–2466, 1999.

27. J. A. K. Suykens, J. De Brabanter, L. Lukas, and J. Vandewalle,

"Weighted least squares support vector machines: robustness and sparce approximation," Neurocomputing, vol. 48, no. 1, pp. 85–105, 2002.

Numerical Well Test Analysis for Polymer Flooding Considering the Non-Newtonian Behavior

Jia Zhichun[1], Li Daolun[1,2], Yang Jinghai[3], Xue Zhenggang[1], and Lu Detang[1]

[1]University of Science and Technology of China, Hefei, Anhui 230026, China
[2]Hefei University of Technology, Hefei, Anhui 230026, China
[3]Daqing Oilfield Company Ltd., Daqing, Heilongjiang 163453, China

ABSTRACT

Well test analysis for polymer flooding is different from traditional well test analysis because of the non-Newtonian properties of underground flow and other mechanisms involved in polymer flooding. Few of the present works have proposed a numerical approach of pressure transient analysis which fully considers the non-Newtonian effect of real polymer solution and interprets the polymer rheology from details of pressure transient response. In this study, a two-phase four-component fully implicit numerical model incorporating shear thinning effect for polymer flooding based on PEBI (Perpendicular Bisection) grid is developed to study transient pressure responses in polymer flooding reservoirs. Parametric studies are conducted to quantify the effect of shear thinning and polymer concentration on the pressure transient response. Results show that shear thinning effect leads to obvious and characteristic nonsmoothness on pressure derivative curves, and the oscillation amplitude of the shear-thinning-induced nonsmoothness is related to the viscosity change decided by shear thinning effect and polymer concentration. Practical applications are carried out with shut-in data obtained in Daqing oil field, which validates our findings. The proposed method and the findings in this paper show significant importance for well test analysis for polymer flooding and the determination of the polymer in situ rheology.

INTRODUCTION

Polymer flooding is one of the most mature EOR (Enhanced Oil Recovery) techniques used in oilfields [1]. It is providing more than 30% of the total oil production in China's Daqing oilfield today. In polymer flooding, polymer solutions injected into the reservoir exhibit non-Newtonian properties [2, 3]. A number of studies have discussed the influence of non-Newtonian behavior of polymer on sweep efficiency and recovery, and the results proved the importance of taking polymer rheology into account for the successful design and evaluation of polymer flooding project [4–6].

The estimation of in situ rheology of polymer solutions in porous medium is difficult in practice [7]. Early time researches in chemical engineering, rheology, and petroleum engineering have focused on the flow behavior of non-Newtonian fluids in porous medium [8–10]. However, the in situ polymer rheology may not consist with the results of those experimental and analytical researches due to the great differences between the actual situation and the simplified and idealized conditions. One approach to directly determine the polymer in situ rheology is by way of well test analysis. The results of several previous works indicated that the well test data from polymer flooding wells can be interpreted to determine the in situ viscosity of non-Newtonian polymer solution [7].

Well test analysis, which is widely used in oilfields, provides valuable information of formation and fluid underground by interpreting the pressure transient response [11]. Yet, conventional well test interpretational models do not work for reservoirs containing non-Newtonian fluids, and well test analysis for polymer flooding is more difficult because of the treatment of polymer rheology and other complicated mechanisms involved in polymer flooding process [11–13].

Researchers have exerted their efforts on well test analysis for non-Newtonian fluid since the 1970s. Ikoku and Ramey [14] and Odeh and Yang [15] proposed mathematical models for flow of non-Newtonian power-law fluid in homogeneous porous medium and induced methods of well test analysis for non-Newtonian fluids. Vongvuthipornchai and Raghavan [16] developed type curves which includes wellbore storage and skin factor for well test analysis for non-Newtonian fluid. Katime-Meindl and Tiab [13] developed the TDS (Tiab's direct synthesis) method for well test analysis and used it for interpretation of non-Newtonian fluid. Recently, Escobar and Martinez et al. [17, 18] applied the TDS technique to pseudo plastic and dilatants fluids in a radial composite reservoir. Yu et al. [19] established a well testing model for polymer flooding based on rheology experiments and presented a numerical well testing interpretation model and analysis techniques to evaluate formation by using pressure transient data in cross-flow double-layer reservoirs. Another recent research of the well test analysis for polymer flooding was performed by Yu et al. [19], presenting a numerical-analytical combined method to infer the in situ polymer rheology from PFO (Pressure Fall-Off) tests.

These works were remarkable and they promoted our understanding of the pressure transient response in polymer flooding. However, despite decades of research, the existing methods of well test analysis for polymer flooding are far from satisfactory for field application. Reasons are as follows: firstly, most of the previous works are analytical approaches, which is mostly confirmed to dealing with certain simplified situations and have limitations in complex field applications. Secondly, the results reported in those papers do not fully describe certain behaviors of actual measured data from polymer flooding wells. Finally, few field examples have been presented which are studied and interpreted to validate the analytical or numerical results.

In this study, a two-phase four-component polymer flooding model is given considering the major physicochemical phenomena including polymer thickening, shear thinning, polymer adsorption, permeability reduction, and dead pore volume. A fully implicit iterative numerical simulator is developed based on PEBI gridding to study the transient pressure response for wells in polymer flooding reservoirs. Parametric studies were conducted to quantify the effect of shear thinning and polymer concentration on the pressure transient response. Finally, a field example was studied in which two pressure build-up tests were performed on the same oil producer from a polymer flooding reservoir in Daqing oilfield. Reservoir parameters and the polymer in situ shear thinning properties were interpreted for the real example, and the impact of adjusting shear thinning on pressure derivative curves is discussed.

METHODOLOGY

Basic Equations for Polymer Flooding

To build the mathematical model for polymer flooding, several assumptions were made to simplify the real situation. The major assumptions in this paper are as follows: (1) the reservoir is isothermal; (2) Darcy's law for fluid flow in porous medium applies; (3) the fluid underground contains two phases: the water phase and the oil phase. The water phase contains three components: pure water, polymer, and salt. (4) The process of polymer adsorption is irreversible. (5) No chemical reaction occurs. (6) Diffusion and dispersion are not considered.

Based on the assumptions above, the mass balance equations were given for each component. Considering the physicochemical phenomena including polymer thickening, polymer adsorption, permeability reduction, and dead pore volume, the basic equations of the polymer flooding model are given by

$$\text{Oil: } \nabla \cdot \left[\frac{KK_{ro}}{\mu_o B_o} (\nabla P_o - \gamma_o \nabla D) \right] + q_o = \frac{\partial}{\partial t} \left(\frac{\phi S_o}{B_o} \right)$$

$$\text{ater: } \nabla \cdot \left[\frac{KK_{rw}}{R_k \mu_w B_w} (\nabla P_w - \gamma_w \nabla D) \right] + q_w = \frac{\partial}{\partial t} \left(\frac{\phi S_w}{B_w} \right)$$

$$\text{polymer: } \nabla \cdot \left[\frac{KK_{rw}}{R_k \mu_w B_w} (\nabla P_w - \gamma_w \nabla D) C_p \right]$$

$$+ q_p = \frac{\partial}{\partial t} \left[\frac{\phi_p \left(S_w C_p + C_{pads} \right)}{B_w} \right]$$

$$\text{salt: } \nabla \cdot \left[\frac{KK_{rw}}{R_k \mu_w B_w} (\nabla P_w - \gamma_w \nabla D) C_{sa} \right]$$

$$+ q_{cl} = \frac{\partial}{\partial t} \left(\frac{\phi S_w C_{sa}}{B_w} \right).$$

$$(1)$$

To enclose the equations, additional relations were induced including the saturation relation and the capillary pressure relation.

Saturation relation:

$$S_o + S_w = 1.$$

$$(2)$$

Capillary pressure relation:

$$P_{cow} = P_o - P_w = P_c \left(S_w, \sigma_{wo} \right).$$

$$(3)$$

The Peaceman model was used to deal with the wells. Let Q be the production of the well at surface condition, considering the wellbore storage, the flow equation in the wellbore is given by

$$\frac{1}{\mu_1 B_1} \frac{2\pi K_{rl} Kh}{\ln (r_{eff}/r_w) + S} \left(p - P_{wf} \right) - \frac{C_{wellbore}}{\Delta t} \left(p_{wf}^{n+1} - p_{wf}^n \right) = Q.$$

$$(4)$$

Equations (1)–(4) are discretized based on PEBI grid by control volume method. And all the unknowns are solved fully implicitly by GMRES matrix solver in this study.

Modeling the Water Phase Viscosity

In the polymer-flooding model, the water phase contains three components: pure water, polymer, and salt. The water phase viscosity is a function of polymer concentration, salinity, and shear rate [2]. The water phase viscosity is calculated following the steps below. Firstly, the unsheared water phase viscosity $\mu 0$ is determined according to the polymer concentration and salinity using

$$\mu_0 = \mu_w \cdot \mu_p,$$

(5)

Where μ_w denotes the viscosity of pure water and μ_p is a viscosity multiplier which is a function of polymer concentration and salinity;

$$\mu_p = \mu_p\left(C_p, C_{sa}\right),$$

(6)

Where C_p is the polymer concentration and C_b is the salt concentration.

In this study, the function relationship represented by (6) is given by experimental data. Namely, a series of μ_p values corresponding to various polymer concentrations are given for different salinities, and μ_p is determined by linearly interpolating in the closed interval from 0 to the maximum given value of C_p and C_{sa}.

Secondly, the effective water phase viscosity is calculated as follows:

$$\mu_{eff} = \mu_w + M \cdot \left(\mu_0 - \mu_w\right),$$

(7)

Where μ_{eff} is the water phase viscosity with shear thinning effect. M is a multiplier representing the shear thinning property of the water phase, which is given by

$$M = \frac{\mu_{\text{eff}} - \mu_w}{\mu_0 - \mu_w},$$

(8)

Where $0<M<1$ which is equivalent to $\mu_w < \mu_{\text{eff}} < \mu_0$.

Similar approaches for the determination of the water phase viscosity have been used by other researchers in numerical studies or commercial simulators for polymer flooding [1, 20].

Modeling the Polymer in Situ Rheology

Most polymers used in EOR are pseudo plastic fluids which exhibits shear thinning behaviors [2, 4]. For pseudo plastic fluids, several common types of rheology models are given as below.

Power-law model is a simple and common used rheology model to describe the non-Newtonian rheology, for which (9) is the constitutive equation and (10) gives the effective viscosity [11]:

$$\tau = K\gamma^n,$$

(9)

$$\mu_{\text{eff}} = K\gamma^{n-1},$$

(10)

Where μ_{eff} is the effective viscosity, K is the consistency index, γ is the shear rate, and n is the power law index.

Another commonly used formula to describe the relationship between viscosity and the shear rate is Meter's equation [8], for which the constitutive equation is

$$\gamma = \frac{\tau}{\mu_\infty + (\mu_0 - \mu_\infty) / \left(1 + (\dot{\gamma}/\dot{\gamma}_{1/2})^{P_\alpha - 1}\right)}$$

(11)

And the effective viscosity is given by

$$\mu_{\text{eff}} = \mu_\infty + \frac{\mu_0 - \mu_\infty}{1 + (\dot{\gamma}/\dot{\gamma}_{1/2})^{P_\alpha - 1}},$$

(12)

Where μ_∞ is the viscosity when shear rate tends to infinity. $\dot{\gamma}_{1/2}$ is the shear rate at which the viscosity equals $(\mu_0 + \mu_\infty)/2$ and P_α is a constant.

Various empirical parameters such as P and $\dot{\gamma}_{1/2}$ are involved if we induce the models of (9)–(12) into the numerical calculation. However, these parameters are difficult to determine in some cases. Therefore, an alternate method which is similar to the treatment of commercial simulators [20] is used in this study, in which the M-V relation is given directly to calculate the non-Newtonian viscosity, where V stands for the flow velocity and M is the shear thinning multiplier defined by (8). The M-V relation indicates the shear thinning properties of non-Newtonian polymer solution.

RESULTS AND DISCUSSION

In this part, a 2000 × 2000 × 10 m single-layer homogeneous reservoir is studied. Two wells are located in the reservoir, as shown in Figure 1. The production well produces at a constant liquid rate of 10 m³/day at surface condition for 30 days. The injection well injects polymer solution at a constant liquid rate of 10 m³/day at surface condition for 20 days and then shuts in for 10 days. Reservoirs parameters for calculation are listed in Table 1. Table 2 gives the relative permeability relation. Table 3 gives the relation between water phase unsheared viscosity (viscosity in the stationary situation) and polymer concentration for salinity equals 3000. Table 3 gives the relationship between the viscosity multiplier μ_p and polymer concentration C_p, which is obtained from experiments on the polymers used in oilfield. The polymer solution is prepared using HPAM of 25 million molecular weight and treated produced water with a total salinity of 3000 mg/L.

Table 1: Reservoir parameters

Parameters	Value
Initial reservoir pressure, MPa	20

Layer thickness, m	10
Horizontal permeability, μm^2	0.5
Porosity	0.2
Compressibility of rock	0.00015
Initial saturation of oil	0.3
Initial saturation of water	0.7
Initial concentration of polymer, kg/m^3	0
Wellbore radius, m	0.1
Liquid rate of production well, m^3/day	10
Liquid rate of injection well, m^3/day	10
Reference pressure for PVT, MPa	20
Viscosity of oil, Pa s	0.01
Viscosity of water, Pa s	0.0006
Volume factor for oil	1.12
Volume factor for water	1.01
Compressibility of oil, 1/MPa	0.0006
Compressibility of water, 1/MPa	0.0006

Table 2: Relation between the water phase saturation and relative permeability

Sw	Krw	Kro
0.3	0	1
0.4	0.02	0.72
0.5	0.05	0.44
0.6	0.09	0.17
0.7	0.15	0.05
0.8	0.24	0

Table 3: Relation between polymer concentration and water phase viscosity

Polymer concentration (kg/m3)	Viscosity multiplier
0	1
0.5	4
1	8
1.5	13
2	21

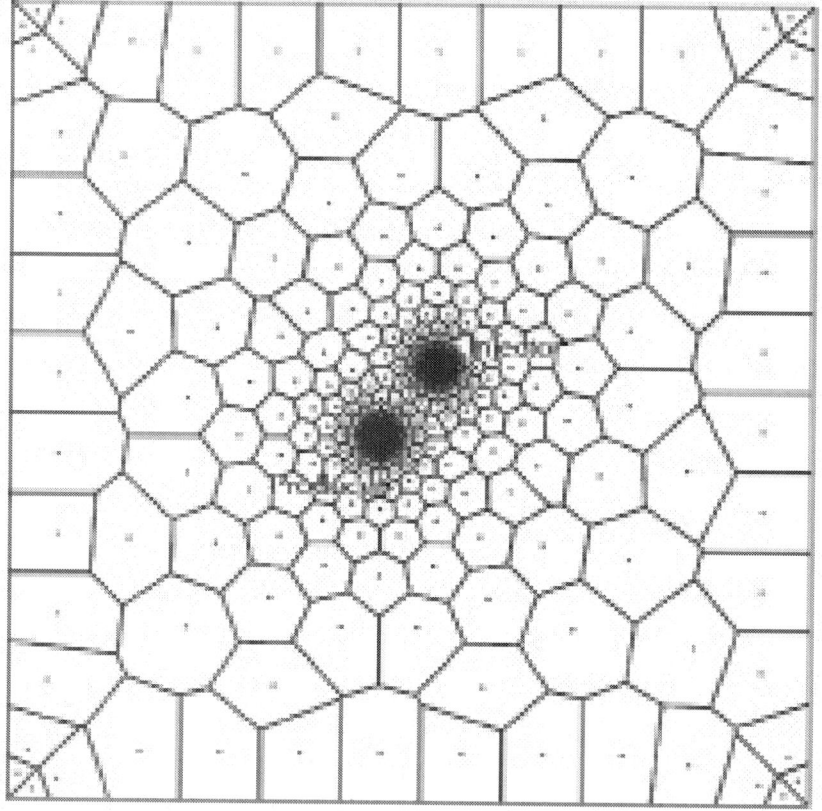

Figure 1: Computational domain and gridding.

Pressure Transient Response for Various Polymer Concentrations without Consideration of the Shear Thinning Effect

In this part, the impact of polymer thickening effect on BHP (bottom-hole pressure) is studied. To eliminate the influent of polymer shear thinning, here we make assumption that the polymer solution behaves as Newtonian type fluid. Figure 2 gives the calculated BHP of the injection well for different injection polymer concentrations. Each of the pressure curves in Figure 2 includes two flow regimes: 0 to 20th day is the injection stage and 20th to 30th day the pressure fall-off stage. Figure 2 shows that for the first flow regime, the injection pressure becomes larger as the polymer concentration increases. This is due to the reduction of flow mobility near the injection well caused by the polymer concentration increase.

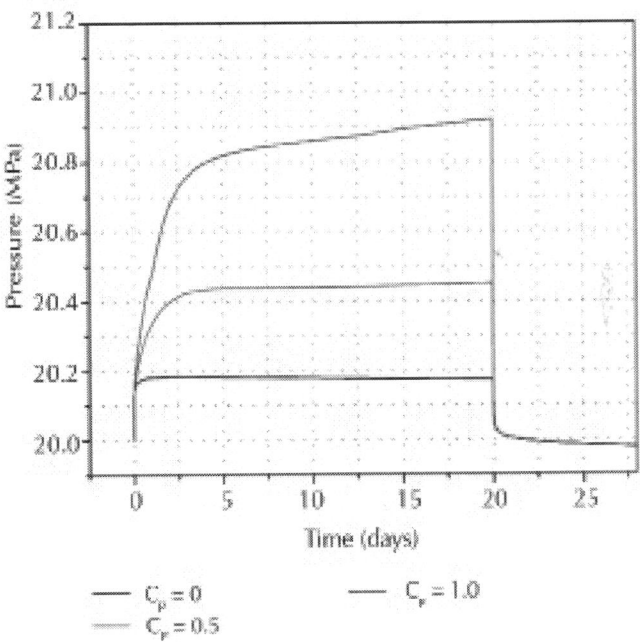

Figure 2: BHP of the injection well for different injection polymer concentrations.

For the pressure fall-off regime, Figure 3 gives the log-log plot of the pressure change and pressure derivative for different injection polymer concentrations. The derivative curves in Figure 3 indicate that the increase of polymer concentration leads to a larger pressure change, which is 0.19673 MPa, 0.47324 MPa, and 0.94508 MPa for polymer concentration equals 0, 0.5, kg/m³, and 1.0 kg/m³, respectively. This is caused by the polymer thickening effect. The growth of injection polymer concentration reduces the mobility of fluid near the injector, thus leading to a larger BHP rise while injecting at a constant liquid rate. Accordingly, after the well shuts in, the pressure drop of the injection well is also larger. Moreover, as the fluid mobility reduces, the flow-continued time of the wellbore storage regime and the transition stage becomes much longer.

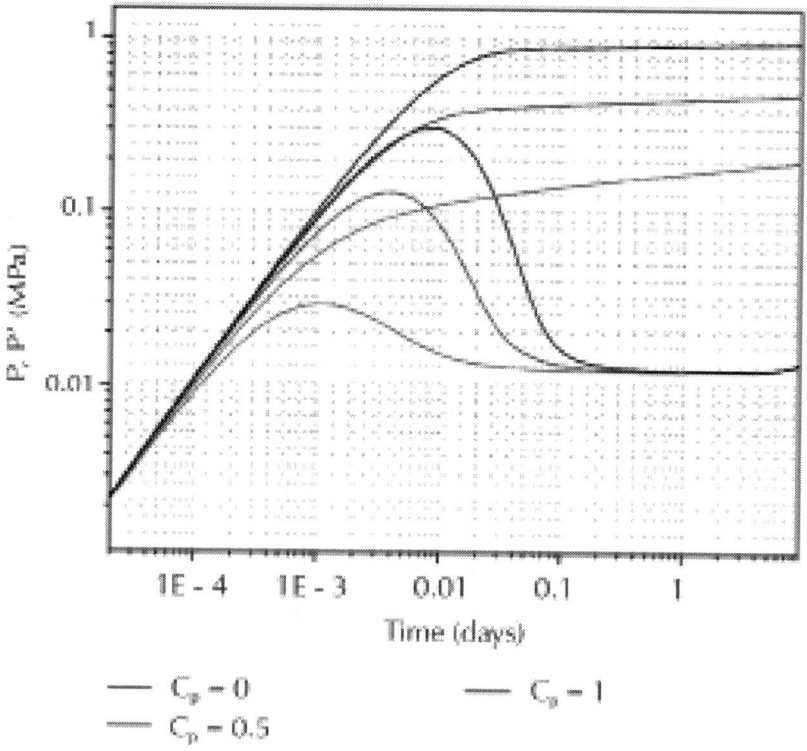

Figure 3: Log-log plot of the well bottom pressure and pressure derivative of the pressure fall-off period for various injection polymer concentrations.

Impact of Shear Thinning Effect on Pressure Transient Response

To analyse the pressure transient response of polymer flooding wells, the shear thinning effect needs to be considered. As is discussed in the previous part, the M-V relation should be given to determine the effective viscosity of water phase with (8). Here, we make assumption that the shear rate of flow in the reservoir is proportional to the water phase velocity, and another simplification is made that the M-V relation is not influenced by the polymer concentration. Three M-V curves used for the calculation in this example are given in Figure 4. It is obvious that the black curve in Figure 4 has the highest rate of descent which indicates a stronger influence of water phase velocity on viscosity. Accordingly, the red curve represents a medium one and the blue curve a small one. The three curves are manually generated according to the following rules: the multiplier M changes from 1 to 0 with the velocity increases. (2) All curves go through the same point (0.1) which means when the velocity is 0, the effective viscosity equals the unsheared viscosity μ_0. (3) As the velocity increases the value of M tends to 0, which means the effective viscosity goes close to the pure water viscosity.

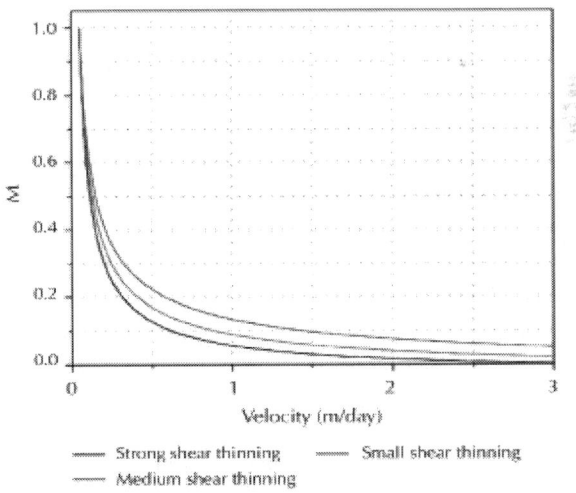

Figure 4: M-V curves.

Figure 5 shows the comparison of BHP calculated under different shear thinning curves and a Newtonian situation, with the same polymer concentration of 1.5 kg/m³. A significant decrease of BHP is showed by the comparison between the results calculated with and without shear thinning, and the BHP calculated under a strong shear thinning is lower than that under a slight one. A qualitative explanation is that shear thinning reduces the effective viscosity of injected polymer solution and makes the injection pressure much lower.

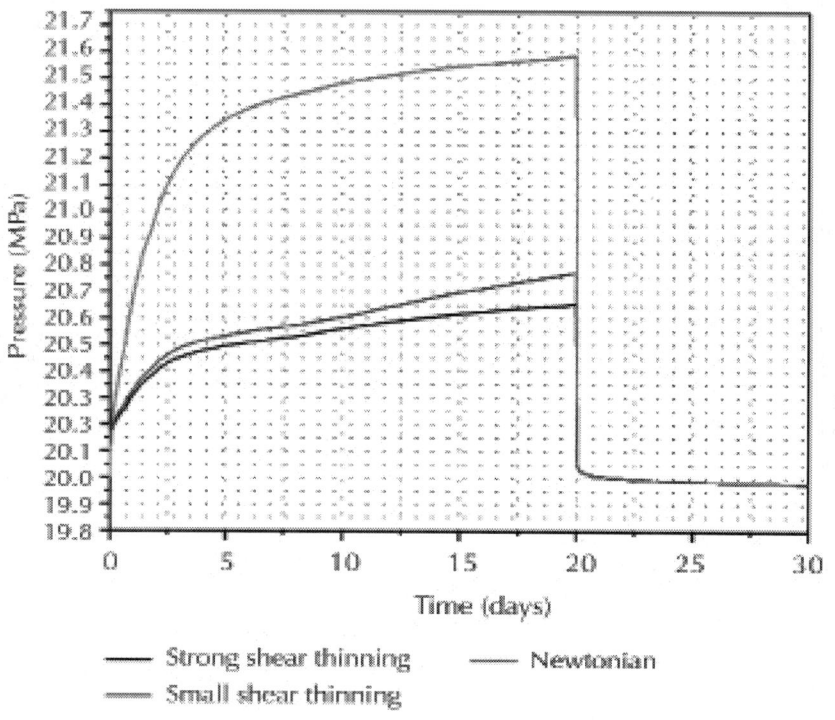

Figure 5: BHP of the injection well for different M-V curves with the same injection polymer concentration of 1.5 kg/m³.

For the fall-off regime, Figure 6 gives the comparison of pressure change and derivative under different shear thinning behaviors and the Newtonian situation. It is showed in the figure that when shear thinning is considered, the pressure change decreases significantly; the stronger the shear thinning is, the smaller the pressure change becomes. This can be explained by considering the impact of shear thinning on the water phase mobility, which is similar to the previous explanation for Figure 3. But we find that Figure 6 differs from Figure 3 in that the flow-continued time of the wellbore storage regime and the transition stage is not significantly influenced by shear thinning. The reason is that the influent of shear thinning is velocity dependent. In the fall-off regime, the sheared viscosity is small at the beginning but increases quickly as the flow slows down until the viscosity value is close to unsheared viscosity. This process will not last long, which makes the flow-continued time under shear thinning still shorter than but not significantly different from that of the Newtonian situation.

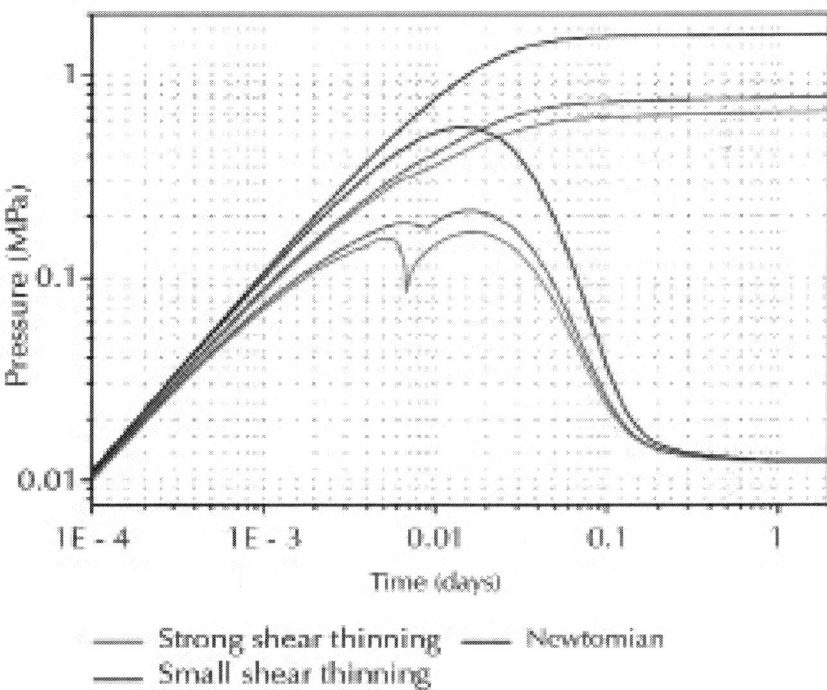

Figure 6: Log-log plot of the well bottom pressure and pressure derivative of the pressure fall-off period for different M-V curves with the same injection polymer concentration of 0.7 kg/m³.

An impressive phenomenon we find from Figure 6 is that the pressure derivative curves are non-smooth in early stage when we consider the shear thinning effect. The non-smoothness of the derivative curves, as we consider, is caused by the rapid change of near-wellbore mobility due to the shear thinning behavior. A qualitative explanation is that when shear thinning exists, the sudden decrease of flow speed after well shuts in causes a dramatic change of water phase viscosity, and this gives a disturbance to BHP, changing the original trend of pressure drop. The disturbance of BHP is more detectable on the derivative which becomes an obvious and characteristic no smoothness at early stage. The nonsmoothness, as caused by the shear thinning behavior, is defined as the shear-thinning-induced nonsmoothness in the paper.

We also find that the oscillation amplitude of the nonsmoothness is affected by relevant factors such as the shear thinning curve and the polymer concentration. Figure 6 shows that the stronger the shear thinning is, the more obvious the nonsmoothness becomes. Figure 7 shows the impact of polymer concentration on pressure transient response with shear thinning effect and the results indicate that the nonsmoothness also becomes more strenuous under a higher polymer concentration.

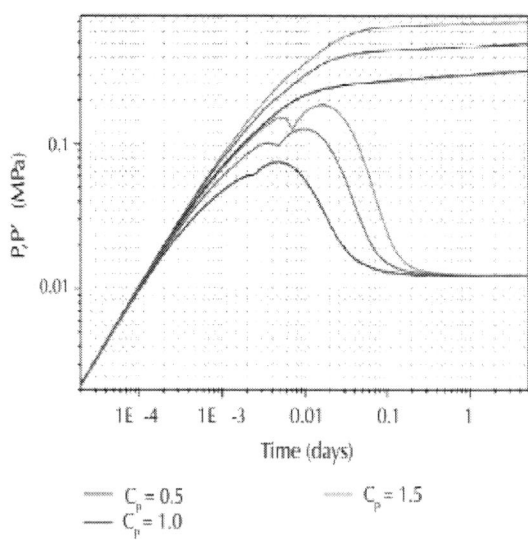

Figure 7: Log-log plot of the well bottom pressure and pressure derivative of the pressure fall-off period for various polymer concentrations under the medium shear thinning.

Other factors we do not consider here may also have influence on the shear-thinning-induced nonsmoothness of derivative curves, which has been observed in field cases, for example, the alternating injection of polymers with different molecular weights. In addition, we do not study the effect of adsorption, inaccessible porosity, and permeability reduction on the transient pressure response in this paper, although they are incorporated in the mathematical model.

It should be noted that consideration of the shear thinning makes the water phase viscosity dependent on velocity, which may greatly increase the difficulty of convergency of the iterative solving and makes the simulation more time consuming.

Transient Pressure Response for Production Well

We also study the pressure transient response of production well in a build-up test in the same example. The injection well injects at a constant liquid rate of $10\,m^3/day$ at surface condition for 30 days. The production well produces solution at the same liquid rate of $10\,m^3/day$ at surface condition for 20 days and then shuts in for 10 days. Reservoirs parameters are the same as those in the previous fall-off case. In order to facilitate the research, we assume that there was a uniform distribution of polymer at the beginning of the simulation. The original polymer concentration of the reservoir is $0.8\,kg/m^3$, and the injected polymer concentration is also $0.8\,kg/m^3$.

Figure 8 gives the comparison of transient pressure responses calculated under the Newtonian and the non-Newtonian situation, respectively. The result of the build-up test indicates similar conclusion to that of a fall-off test: shear thinning reduces the pressure change and draws down the peak value of derivative and induces early stage nonsmoothness on the derivative curve. As there exists an original distribution of polymer, the curves in Figure 8 are different from those in Figure 6 in late stage.

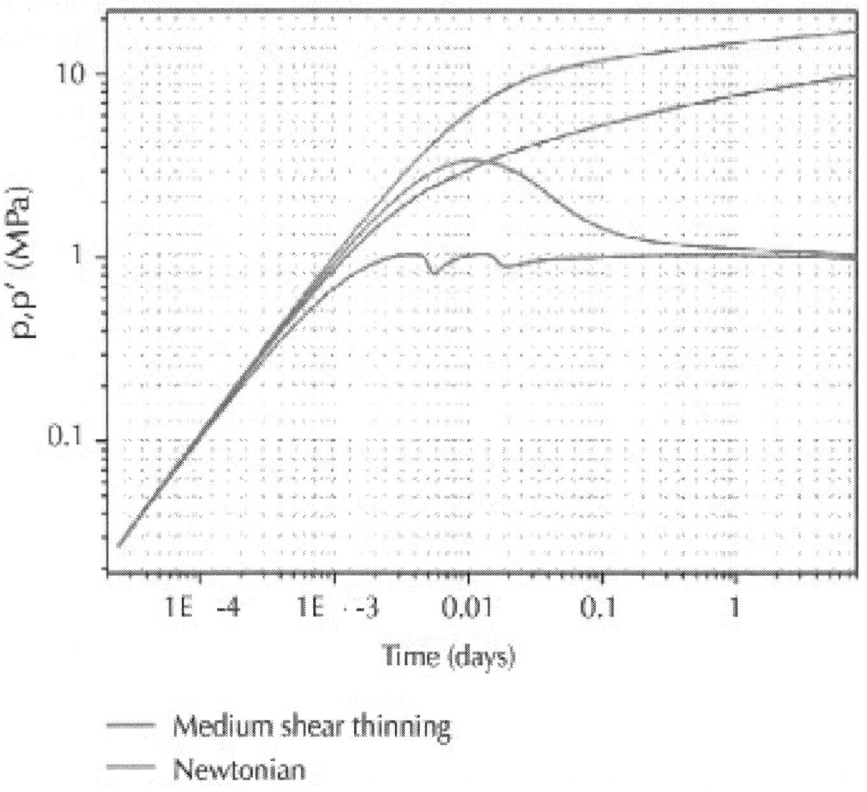

— Medium shear thinning
— Newtonian

Figure 8: Log-log plot of the well bottom pressure and pressure derivative of the production well for different M-V curves.

CASE STUDY

We study a five-well group from a polymer flooding reservoir in Daqing oil field with a production well in the center, and other four injection wells around it. The average layer thickness is 5.5 m. The positions of the wells are shown in Figure 9. Water flooding production lasts for years in the reservoir before the polymer flooding stage, and polymer injection started in October 2009.

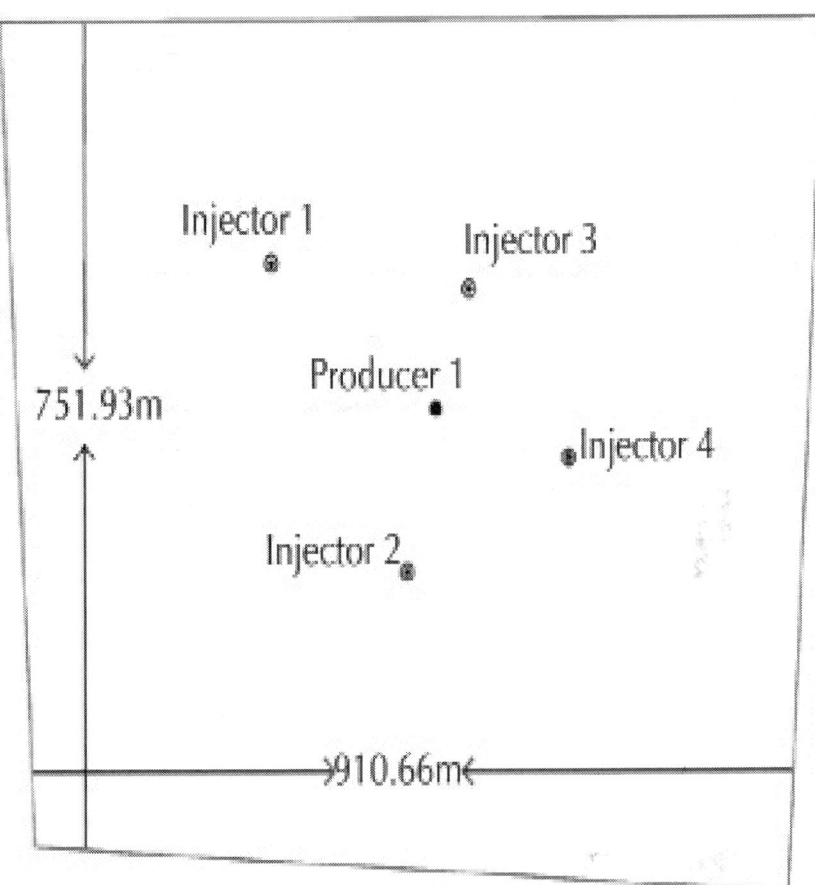

Figure 9: Computational domain for the real example.

Two pressure build-up tests were performed for the oil producer in 2010.04 and 2012.04, respectively, for which the details are given in Table 4. Log-log plots of the pressure change and pressure derivative curves for the two well tests are demonstrated in Figures 11 and 12.

Table 4: Details of the well tests

Name	Test1	Test2
Date of test	2010/04/19	2012/04/15
Well	Producer 1	Producer 1
Well test method	Pressure build-up	Pressure build-up
Time of stable production, day	150	150
Liquid rate of stable production, m³/day	74	30
Time of well shutdown, day	3.01	3.09
Measured concentration of polymer in the production well, kg/m³	0.009	0.85

We can see from Figures 10 and 11 that pressure transient response of the two build-up tests is different in that the derivative curves of Test1 is smooth while the derivative curve of Test2 shows early stage nonsmoothness. We explain this with the shear-thinning-induced nonsmoothness. Here are some evidences. By referring to the record of production, we know that polymer injection began in 2009.10. Polymer concentration detected at the production well stays very low at first until a marked increase in 2010.07, and polymer concentration by the time of 2010.04 is only 0.009 kg/m³ which means the injected polymer solution had not reached the production well in 2010.4. Therefore, pressure derivative curve of Test1 is not affected by the polymer shear thinning and shows typical characteristic of Newtonian flow situation. While for Test2, by the time of which the measured polymer concentration at the production well is 0.85 kg/m³. So according to the discussion of the example given above, it is possible that shear thinning of polymer solution affects the derivative curve of Test2, making it nonsmooth in the early stage.

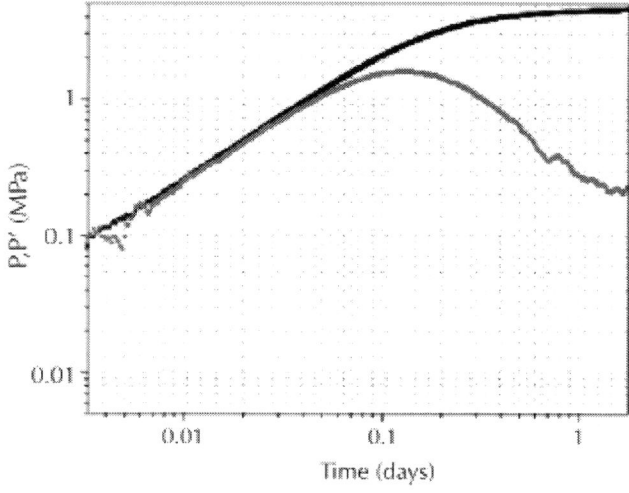

* Pressure change
· Pressure derivative

Figure 10: Measured pressure change and pressured derivative curves for Test1.

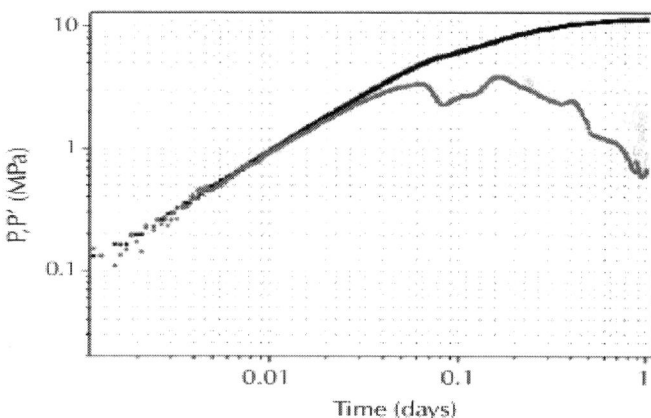

· Pressure change
· Pressure derivative

Figure 11: Measured pressure change and pressured derivative curves for Test2.

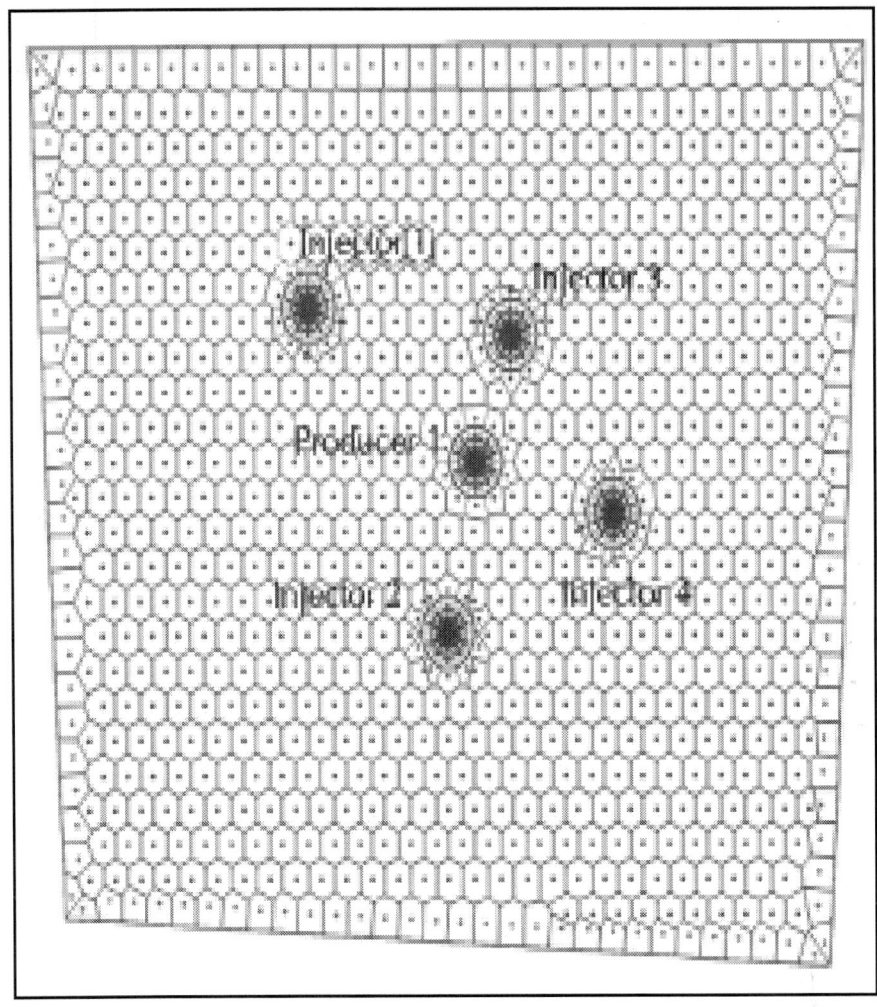

Figure 12: Gridding for the real example.

We design numerical calculation and perform well test interpretation for the measured transient pressure data. The calculation domain and gridding are shown in Figure 12, and other parameters for calculation are given in Table 5. The matched results of the measured pressure data and numerical calculation data are shown in Figures 13 and 14. The interpreted shear thinning curve is shown in Figure 15, and other interpreted reservoir parameters are given in Table 6.

Table 5: Parameters for calculation

Parameters	Value
Layer thickness, m	5.5
Horizontal permeability, μm2	Permeability distribution
Porosity (initial)	0.2
Compressibility of rock, 1/MPa	0.00015
Initial saturation of oil	0.35
Initial saturation of water	0.65
Wellbore radius, m	0.1
Reference pressure for PVT Calculation, MPa	2
Viscosity of oil, Pa s	0.0067
Viscosity of water, Pa s	0.0006
Volume factor for oil	1.12
Volume factor for water	1
Compressibility of oil, 1/MPa	0.00084
Compressibility of water, 1/MPa	0.000449

Table 6: Interpreted parameters

Date	Near-wellbore permeability, μm2	Wellbore storage, m3/MPa	Skin	Porosity
2010.4	0.35	2.65	4.8	0.2
2012.4	0.25	0.6	9	0.2

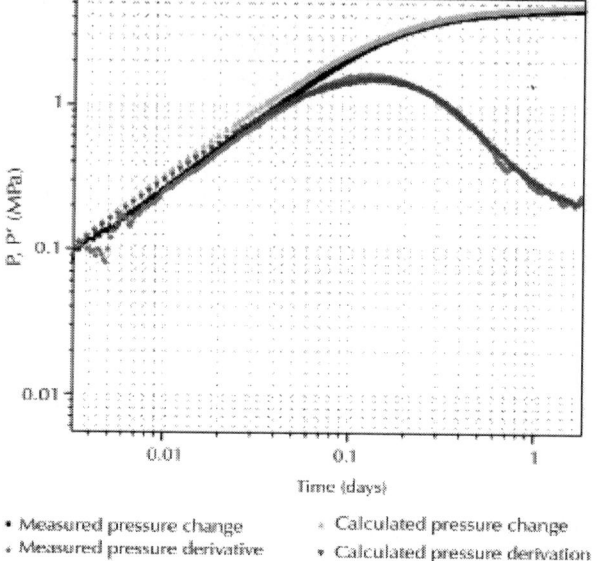

Figure 13: Pressure change and pressure derivative history match for Test1.

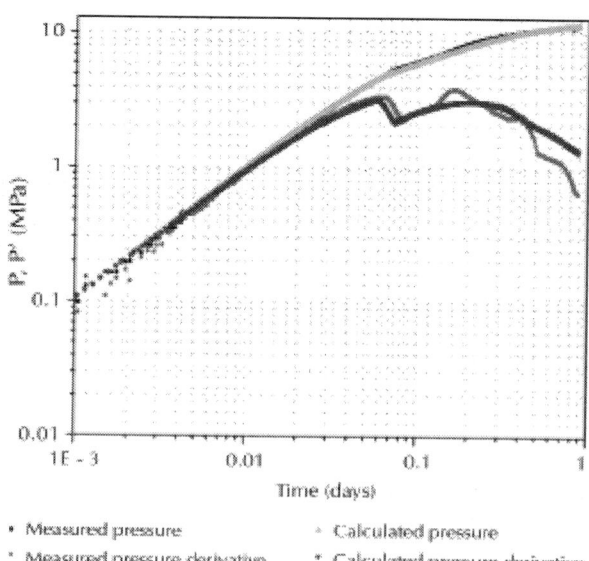

Figure 14: Pressure change and pressure derivative history match for Test2.

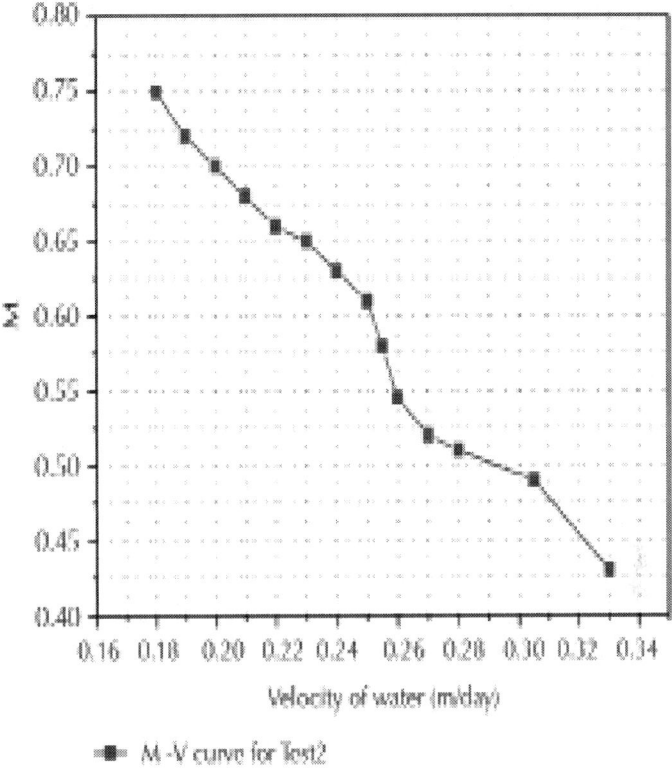

Figure 15: Interpreted - shear thinning curve.

The result of numerical simulation indicates that by the beginning of well test 1, polymer concentration in the production well is 0, which means that injected polymer had not reached the production well. Figure 13shows that the measured pressure change and the derivative curve of the pressure are well matched by the numerical result. We note that the calculated derivative curves of Test1 are smooth in early stage. For Test2, numerical result shows polymer concentration of the producer is $0.74\,kg/m^3$, while the measured data is $0.85\,kg/m^3$. The results indicate that polymer solutions injected into the reservoir had already reached the oil producer by the time of Test2. Figure 14 shows that not only the whole shape of the curves is well matched but also the nonsmoothness of the derivative is fitted through the interpretation. During the fitting for Test2 we find that the oscillation amplitude and position of the nonsmoothness are quite sensitive to the shear thinning curves, and small change of the shear thinning curve may lead to obvious change of the nonsmoothness. This indicates that in well test analysis the nonsmoothness of the derivative curve is a significative feature which can be interpreted to determine the shear thinning curve.

The finding shows significant importance for practical well test analysis for the following reasons: first, the nonsmoothness of pressure derivative curves is a common phenomenon of well test for polymer flooding in oilfields, and it is mostly considered to be caused by the heterogeneity of polymer slug, or measurement error. The finding of this study provides a new explanation for the frequently occurring of derivative curve nonsmoothness. Second, while most shear thinning data used in oil field are obtained from simplified experiments and do not represent the real in situ rheology, this finding provides an approach to estimate the polymer in situ rheology from transient pressure data.

Other interpreted parameters in Table 6 show that the near-wellbore permeability decreased and skin increased from Test1 to Test2. The interpreted data in Table 6 reflect the damage of the formation and wellbore by polymer solution.

CONCLUSIONS

In this paper, a polymer flooding model is presented and a fully implicit numerical simulator was developed based on the PEBI grid. Using the numerical simulator, parametric studies are conducted to study the pressure transient analysis for polymer flooding. Finally, a field case is studied and numerical well test analysis is performed to interpret the measured data.

By performing parametric studies, we find that the polymer concentration and polymer shear thinning effect greatly affect the BHP. The pressure transient response under different polymer concentration without consideration of the shear thinning effect show that the increasing polymer concentration makes a larger pressure change, and a longer flow-continued time of the wellbore storage stage and the transition stage. By considering the shear thinning effect, we find that the polymer shear thinning leads to a smaller pressure change, and the flow-continued time of the early stage is not significantly affected by shear thinning effect. Shear thinning also makes the pressure derivative curve nonsmooth during the wellbore storage and the transition stages, which is caused by the rapid change of the fluid mobility. The oscillation amplitude of the nonsmoothness is influenced by the M- V shear thinning curve and the polymer concentration.

A field case is studied in which two pressure build-up tests were performed on an oil producer in different stages of polymer flooding process. Significant differences are detected between the measured pressure curves of the two well tests, among which the pressure derivative curve of the later one showed noticeable feature of polymer shear thinning while the derivative curve of the earlier well test does not. The measured pressure data validate our parametric study. Numerical calculation and well test interpretation are also performed for the field example. It is found that the shear thinning calculation has a significant impact on fitting of the transient pressure curves, and the nonsmoothness of pressure derivative is possibe to be interpreted to determine the polymer in situ non-Newtonian properties.

ACKNOWLEDGMENTS

This work was sponsored by Major State Basic Research Development Program of China (973 Program) (no. 2011CB707305), National Key Science and Technology Project (2011ZX05009-006), and CAS Strategic Priority Research Program (XDB10030402).

REFERENCES

1. A. M. AlSofi and M. J. Blunt, "Streamline-based simulation of non-Newtonian polymer flooding,"SPE Journal, vol. 5, no. 4, pp. 895–905, 2010.

2. L. W. Lake, Enhanced Oil Recovery, Prentice Hall, Englewood Cliffs, NJ, USA, 1989.

3. S. Vongvuthipornchai and R. Raghavan, "Pressure falloff behavior in vertically fractured wells: non-newtonian power-law fluids. spe formation evaluation," SPE Formation Evaluation, vol. 2, no. 4, pp. 573–589, 1987.

4. A. M. AlSofi, T. C. LaForce, and M. J. Blunt, "Sweep impairment due to polymers shear thinning," inProceedings of the 16th Middle East Oil and Gas Show and Conference (MEOS '09), pp. 834–845, Bahrain, Bahrain, March 2009.

5.   X. Lopez and M. J. Blunt, "Predicting the impact of non-newtonian rheology on relative permeability using pore-scale modeling," in Proceedings of the SPE Annual Technical Conference and Exhibition, Paper SPE 89981, pp. 793–800, Houston, Tex, USA, September 2004.

6.   B. Wang, L. W. Lake, and G. A. Pope, "Development and application of a streamline micellar/polymer simulator," in Proceedings of the SPE Annual Technical Conference and Exhibition, Paper SPE10290, San Antonio, Tex, USA, October 1981.

7.   A. Laoroongroj, M. Zechner, T. Clemens, and A. Gringarten, "Determination of the in-situ polymer viscosity from fall off tests," in Proceedings of the SPE Europec/EAGE Annual Conference, Paper SPE 154832, Copenhagen, Denmark, June 2012.

8.   D. M. Meter and B. R. Bird, "Tube flow of non-Newtonian polymer solutions, parts 1 and 2-laminar flow and rheological models," American Institute of Chemical Engineer, vol. 878–881, pp. 1143–1150, 1964.

9.   J. G. Savins, "Non-newtonian flow through porous media," Industrial & Engineering Chemistry, vol. 61, no. 10, pp. 18–47, 1969.

10.   W. B. Gogarty, G. L. Levy, and V. G. Fox, "Viscoelastic effects in polymer flow through porous media," in Proceedings of the Fall Meeting of the Society of Petroleum Engineers of AIME, Paper SPE 4025, San Antonio, Tex, USA, October 1972.

11.   X. Y. Kong, Transport in Porous Media, Printing House of University of Science and Technology of China, Hefei, China, 2010 (Chinese).

12.   C. Huh and T. Snow, "Well testing with a non-Newtonian fluid in the reservoir," in Presented at the SPE Annual Technical Conference and Exhibition, Las Vegas, Nev, USA, September 1985, Paper SPE14453.

13.   I. Katime-Meindl and D. Tiab, "Analysis of pressure transient test of non-newtonian fluids in infinite reservoir and in the presence of a single linear boundary by the direct synthesis technique," inProceedings of the SPE Annual Technical Conference and Exhibition, Paper SPE 71587, pp. 2213–2222, New Orleans, Lo, USA, October 2001.

14. C. U. Ikoku and H. J. Ramey Jr., "Transient flow of non-Newtonian power-law fluids in porous medium," SPE Journal, vol. 19, no. 3, pp. 164–174, 1979, SPE 7139-PA.

15. A. S. Odeh and H. T. Yang, "Flow of non-newtonian power-law fluids through in porous medium,"SPE Journal, vol. 19, no. 3, pp. 155–163, 1979, SPE 7150-PA.

16. S. Vongvuthipornchai and R. Raghavan, "Well test analysis of data dominated by storage and skin: non-Newtonian power-law fluids," SPE Formation Evaluation, vol. 2, no. 4, pp. 618–628, 1987, SPE14454-PA.

17. F.-H. Escobar, J.-A. Martinez, and M. Montealegre-Madero, "Pressure and pressure derivative analysis for a well in a radial composite reservoir with a non-newtonian /newtonian interface," Ciencia, Tecnología y Futuro, vol. 4, no. 1, pp. 33–42, 2010.

18. J. A. Martinez, F. H. Escobar, and M. Montealegre, "Vertical well pressure and pressure derivative analysis for bingham fluids in a homogeneous reservoirs," Dyna, vol. 78, no. 166, pp. 21–28, 2011.

19. H. Y. Yu, H. Guo, Y. W. He et al., "Numerical well testing interpretation model and applications in crossflow double-layer reservoirs by polymer flooding," The Scientific World Journal, vol. 2014, Article ID 890874, 11 pages, 2014.

20. H. Mahani, T. G. Sorop, P. J. van den Hoek, A. D. Brooks, and M. Zwaan, "Injection fall-off analysis of polymer flooding EOR," in Proceedings of the SPE Reservoir Characterisation and Simulation Conference and Exhibition, Paper SPE 145125, Abu Dhabi, UAE, October 2011.

J

Chapter 4

Electrolytic Treatment of Wastewater in the Oil Industry

Alexandre Andrade Cerqueira
and Monica Regina da Costa Marques

[1]Instituto de Química – Laboratório de Tecnologia Ambiental (LABTAM), Programa de Pós-Graduação em Meio Ambiente (PPG-MA), Universidade do Estado do Rio de Janeiro (UERJ),, Brazil

INTRODUCTION

Industrial development in recent decades has been a major contributor to the degradation of water quality, both through negligence in treatment of wastewater before discharge into receiving bodies and accidental pollutant spills in aquatic environments [1].

The importance of oil to society is unquestioned. It is not only a major source of energy used by mankind, but its refined products are the raw material for the manufacture of many consumer goods [2].

A world without the amenities and benefits offered by oil would require a total change of mindset and habits among the population, a total overhaul of the way our society works. At the same time, the oil industry is a major source of pollutants that degrade the environment, with the potential to affect it at all levels: air, water, soil, and consequently, all living beings on the planet [2].

Oil and its derivatives are the most important pollutants, due to, among other factors, the increasing amounts that have been extracted and processed. Also, carelessness and neglect of safety standards and routine maintenance of equipment (pipelines, terminals, and platforms) aggravate the water pollution problems caused by the oil industry [3].

Due to the negative environmental impacts of exploration and production of oil, new more restrictive environmental laws and regulations have been issued. It is estimated that in the United States alone, the oil industry will need to invest about 160 billion dollars in actions to protect the environment over the next 20 years to meet environmental legislation more demanding than currently adopted in Brazil [4].

One of the crucial points to be attacked is the issue of water production, which is generated in this activity, which is increasing in volume as they get older wells and new wells are drilled [5]. On average for each m^3/day of oil produced, 3-4 m^3/day of water is produced, although this figure can reach up to 7 m^3/day or even more in exploration, drilling and production. The water produced along with oil corresponds to 98% of the effluents. It contains salts, oils and other toxic chemicals in addition to having high temperature and no oxygen [6].

According to [7], treatment of produced water is an urgent matter in view of the high daily volume. Different processes have been described for the treatment of such effluent, but the most frequently used are chemical destabilization [8,9] and electrochemical destabilization [10,11]. Biological processes are rarely used since these effluents usually contain biocides [12].

The use of EF can enable the release into receiving bodies or reinjection in wells of the treated effluent by reducing the organic load and removing oily and solid particles in suspension [13].

According to [14], the current EF technology inherently involves the formation of an impermeable oxide layer on the cathode and deterioration of the anode due to oxidation. This leads to loss of efficiency of the electro flocculation unit. These limitations of the process have been decreased to some extent by the addition of parallel plate electrodes in the cell configuration. However, the use of alternating current in EF retards the normal mechanisms of electrode deterioration that are inherent in DC system due to cyclical energization, thus increasing the electrode life.

In the present text, we evaluate the efficiency of electro flocculation with direct current and variable frequency alternating current with the use of aluminum electrodes for the treatment of oily wastewater from actual production.

Petroleum Exploration and Production

Petroleum is the name given to natural mixtures of hydrocarbons, which can be found in the solid, liquid or gaseous state depending on the conditions of temperature and pressure [15].

Oil is a combination of carbon and hydrogen molecules and is less dense than water, with a characteristic odor and color varying from black to brown. Although the subject of much discussion in the past, today oil's organic origin is accepted. Oil exploration and production is one of the most important industrial activities of modern society and its derivatives have many industrial applications. Because of the need to meet the growing demand for the product, the extraction of oil has increased greatly in recent decades. However, this extraction to meet world oil demand causes damage to the environment, with the main culprit being produced water [16,17].

Oily Water

Oily water is a generic term used to describe all water which contains varying amounts of oils and greases in addition to a variety of other materials in suspension. These can include sand, clay and other materials, along with a range of dissolved colloidal substances, such as detergents, salts, metal ions, etc. To meet environmental standards for disposal and/or the characteristics necessary for reuse, the treatment of oily water can be complex, dependent on highly efficient processes.

In the petroleum industry, oily water occurs in the stages of production, transportation and refining, as well as during the use of derivatives. However, the production phase is the largest source of this pollution. During the production process, oil is commonly extracted along with water and gas. The associated water can reach

50% of the volume produced, or even approaching 100% at the end of the productive life of wells. The discharge or reinjection of this co-produced water is only permitted after removal of oil and suspended solids to acceptable levels [18].

The terms "produced water," "petroleum water", "formation water" and "oily water" are used to refer to the water extracted along with oil [17,19].

The composition of this produced water is very complex. Depending on its origin it can contain a wide variety of chemicals such as organic salts, aliphatic and aromatic hydrocarbons, oils and greases, metals, and occasionally radioactive materials. A striking feature of the water coming from offshore oil is its high salinity [17, 20, 21], which expressed as chloride ions (Cl⁻) can reach 120 g/L [22].

In oil wells under the seabed, the amount of this wastewater can reach 90% of all effluent during the production of oil and can be 7-10 times higher than the oil extracted from a given well [17, 21].

A new oil field produces little oily water (about 5-15% of the total oil produced). However, as the well becomes exhausted, the water volume can increase significantly, to the range 75-90%. This excessive production of water has become a major concern in the oil and gas industry [23]. Before disposal into receiving bodies or use for re-injection into wells, it is necessary to treat this water because the large amounts of pollutants cannot be discharged into the marine environment [24].

Electrocoagulation

EC is a process that involves the generation of coagulants "in situ" from an electrode by the action of electric current applied to these electrodes. This generation of ions is followed by electrophoretic concentration of particles around the anode. The ions are attracted by the colloidal particles, neutralizing their charge and allowing their coagulation. The hydrogen gas released from the cathode interacts with the particles causing flocculation, allowing the unwanted material to rise and be removed (Figure 1). Various metals have been tested as electrodes, such as aluminum, iron, stainless steel and platinum [25].

The theory of EC has been discussed by several authors, and depending on the complexity of the phenomena involved can be summarized in three successive stages of operation:

- Formation of a coagulating agent through the electrolytic oxidation of the sacrificial electrode, which neutralizes the surface charge, destabilizes the colloidal particles and breaks down emulsions (coagulation – EC step);

- the particle agglutination promoted by the coagulating agent facilitates the formation and growth of flakes (flocculation – EF step) and,

- generation of micro-bubbles of oxygen (O_2) at the anode and hydrogen (H_2) at the cathode, which rise to the surface and are adsorbed when colliding with the flakes, carrying the particles and impurities in suspension to the top and thereby promoting the clarification of the effluent (flotation – electroflotation step).

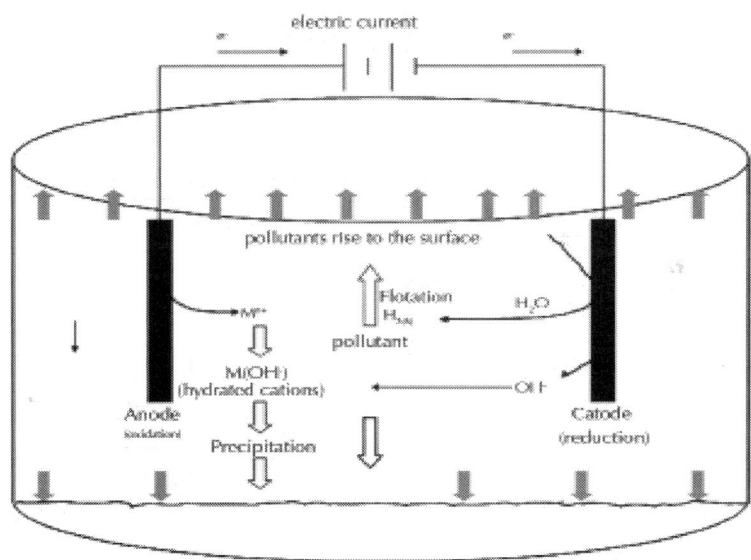

Figure 1: Schematic diagram of an electrocoagulation cell with two electrodes. Source: Adapted from [26].

Processes for electrochemical treatment of effluents have been described in the literature since 1903. In recent years interest has been growing, especially because of its simplicity of operation and application to treat various types of effluents from various sectors, such as domestic sewage [27], laundries [29], restaurants [30] steel mills [31], textile mills [32], and tanneries [33], facilitating the removal of metal ions [28], fluoride ion [34], boron [35] and oils [7, 36-41].

Several types of reactors have been proposed in the literature: mono-polar, bipolar etc. But the most widely used is the monopolar reactor [14]. In its simplest form, a monopolar EF reactor is composed of an electrolytic cell with an anode and a cathode. In this case, large-area electrodes must be used, or electrodes connected in parallel. In the parallel arrangement, the electric current is divided among all the electrodes in relation to the resistance of individual cells. Thus, a lower potential difference is required in connection of this type when compared to a series arrangement.

For electrodes in series, a higher potential difference is required for a given current flow, because the electrodes are connected in series and have a higher resistance. The same current, however, runs through all the electrodes, and the current is divided among all the individual electrodes of the cells [14].

In the case of the bipolar reactor, the sacrifice electrodes are placed between the two electrodes in parallel (called conductive plates), without any electrical connection. Only two monopolar electrodes are connected to the power source, with no interconnection between the sacrifice electrodes. When the current passes through the two parallel electrodes, the neutral sides of the plate acquire an opposite charge than monopole electrode. The external electrodes are monopolar and the internal ones are bipolar.

According to [42], most of the setups for treatment of effluents, the electrodes are made of identical material, mainly due to the following reasons:

- equal electrodes, made of the same material, have the same electrode potential;
- electrodes of different materials imply the use of materials other than iron or aluminum, which increases the cost;
- electrodes of the same material suffer the same wear, which simplifies their replacement.

In any electrochemical process, the electrode material has a significant effect on the effluent treatment. For the treatment of drinking water, it should be nontoxic, have low cost and be readily available [31].

Generally, however, iron electrodes have the disadvantage that the effluent has a pronounced green or yellow color during and after treatment. This coloration comes from the Fe^{2+} (green) and Fe^{3+} (yellow) generated in the electrolytic treatment. In contrast, with aluminum electrodes the final effluent is clear and stable, with no residual coloring.

In the work presented by [43], when aluminum and iron electrodes were tested under the same conditions, using direct current, the results for COD, turbidity and suspended solids were better for the aluminum than the iron electrodes. This advantage was also observed by [30]. However, when comparing the removal of arsenic by iron and aluminum electrodes, [31] found that the iron electrode was better because it showed 99% removal to 37% for aluminum. This difference was explained because the adsorption capacity of the $Al(OH)_3+$ by As^3 is much smaller than that of $Fe(OH)_3$.

Tests carried out by [44] of COD, phenols and turbidity of hydrocarbons from a petrochemical plant, using iron and aluminum electrodes, showed better performance by aluminum electrodes.

According to [45], just as in electrocoagulation, the removal of pollutants closely depends on the size of the bubbles generated, while energy consumption is related to the electrolytic cell design, electrode materials, arrangement of electrodes and operating conditions, such as current density, conductivity of the effluent and electrolysis time, among others. The difference in size of the bubbles in the effluent depends on the pH, current density, electrode material and surface condition of the electrodes.

The mechanism of EC is highly dependent on the chemistry of the aqueous medium, especially conductivity. Moreover, other characteristics such as pH, particle size and concentration of the constituents will influence the electrocoagulation process [14].

In an EC reactor, the rate of coagulant addition is determined by the kinetics of the electrodes. The reactions at the electrodes are heterogeneous and take place in the interfacial region between the electrodes and the solution. Since the reaction consists of electron transfer via an interface, this reaction will be influenced by the characteristics of this interface, such as the potential difference that is established in equilibrium and changes in potential across the interface in function of distance.

The potential of the electrolysis is strongly dependent on current density, effluent conductivity, distance between the electrodes and the surface condition of the electrodes.

Parameters Influencing Electrolytic Processes

pH

The EC process performance is greatly influenced by the pH of the solution [46]. Considering only mononuclear speciation, the total aluminum present in solution (α) at a given pH value can be calculated (Figure 2). This distribution diagram shows the extent of hydrolysis, which depends on the total metal concentration and pH. As the pH increases, the dominant species changes, in this case from the Al^{3+} cation to the $Al(OH)_4^-$ ion not participate in the coagulation reactions and tend to remain in solution [47].

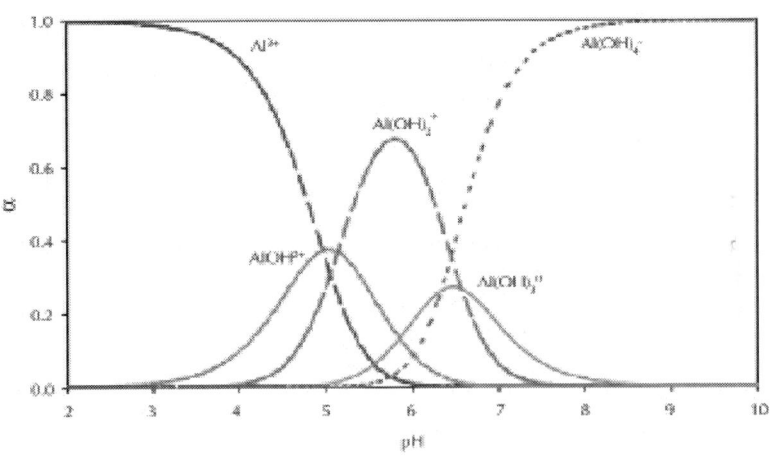

Figure 2: Diagram of distribution for Al-H$_2$O considering only mononuclear species (Source: [25]).

The solubility diagram for aluminum hydroxide, $Al(OH)_3$, is shown in Figure 3. The solubility boundary denotes the thermodynamic equilibrium that exists between the dominant aluminum species in solution at a given pH and the solid aluminum hydroxide. The minimum solubility (0.03 mg Al/L) occurs at a pH of 6.3 and increases as the solution becomes more alkaline or acid [48].

Thus, the active metal cations produced in the anode react with the OH⁻ ions produced at the cathode to form a metal hydroxide, which then acts as a coagulant with the polluting particles and the metal hydroxides, forming larger aggregates, which can both undergo sedimentation and be carried to the surface of the hydrogen bubbles generated at the cathode.

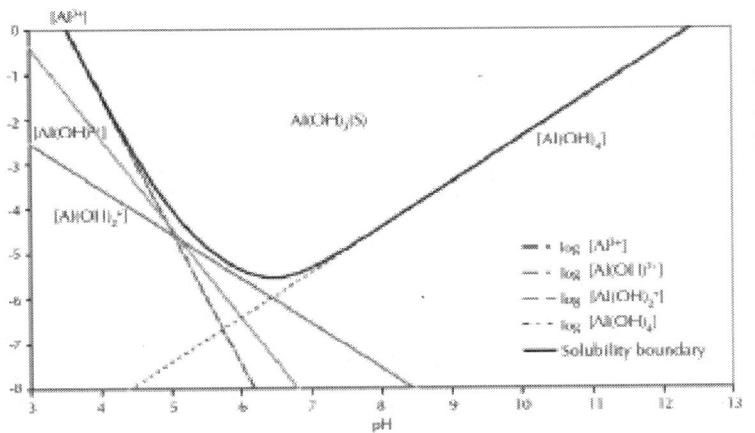

Figure 3: Solubility diagram of aluminum hydroxide Al (OH)₃ considering only mononuclear species of Al (Source: [25]).

The distribution and solubility diagrams presented above consider only mononuclear aluminum species, whereas in reality this system is considerably more complex. As the aluminum concentration increases, polynuclear complexes of aluminum can be formed and the aluminum hydroxide is precipitated, as illustrated by equation (1):

$$Al^{3+}\ Al(OH)^{3-n}{}_n\ Al_2(OH)^{4+}{}_2\ Al(OH)^{5+}{}_4\ Al_{13}\ complex\ Al(OH)_3 \tag{1}$$

By studying a continuous electrocoagulation process using aluminum electrodes and varying the pH, [7] observed that in acidic (pH <4) or alkaline (pH> 9) media, in which cationic or anionic monomeric species of aluminum are predominant, the emulsion remained stable and there was no decrease in COD. Moreover, at pH 5 to 9, the removal was 80%. Under these conditions (pH 5 to 9), the predominant species are polymeric complexes of aluminum and amorphous precipitate of aluminum hydroxide. The surface of the latter can be positively or negatively charged by adsorption of ions from the solution.

Distance between Electrodes

The distance between the electrodes is an important variable to optimize operating costs. According to [49], when the effluent conductivity is relatively high, a bigger space between the electrodes should be used. In contrast, in situations of moderate value, it is recommended to use a smaller distance, as this will reduce power consumption without changing the degree of separation, because in this case the current would not be altered.

When testing a treatment system under the same electric current, [50] observed there was no difference in removal efficiency for different spacings between the electrodes. Therefore, the distance between them is considered to be only a factor for cost optimization.

In turn, [51] reported in their paper that with increasing distance between the electrodes, fewer interactions of ions of the solution with the coagulant will occur. The difference in conclusions between the two research teams can be attributed to a possible divergence in the conductivity value of each effluent, since if it was high in the first study (range between 100 and 140 mS cm^{-1}), there would not have been any change in removal efficiency, because even with a greater distance between the electrodes, there would be a minimum conductivity of the solution that would carry the current.

In the second study, the authors did not mention the effluent's conductivity value. However, it can be assumed it was lower than in the first study, because increasing the distance between the electrodes caused the interactions to decrease and there would need to be a minimum conductivity to ensure the transmission of electric current. Therefore, for there to be no difference in removal with changes in the spacing between electrodes, the treated solution must have a minimum electrical conductivity value.

The greater the distance between the electrodes, the greater must be the voltage applied, because the solution has resistance to the passage of electric current. Thus, according to the characteristics of the effluent, the distance between electrodes can be varied to maximize the efficiency of the process. For example, longer distances can be used when the effluent conductivity is relatively high, while the distance should be as small as possible when conductivity is low so that does not overly increase the need for power.

Electrical Conductivity of the Effluent

The increase in conductivity by addition of sodium chloride is known to reduce the cell voltage due to the reduction of the ohmic resistance of the effluent [43, 52]. Chloride ions can significantly reduce the adverse effects of other anions such as HCO_3^- and SO_4^{2-}.

The electrical conductivity of the effluent is a variable that affects the current efficiency, the cell voltage and power consumption. It is also important when optimizing the parameters of the system, since high conductivity associated with a small distance between electrodes minimizes the consumption of energy, but does not affect the efficiency of removing contaminants, as shown in [53].

When the electric conductivity of an effluent is too low, sodium chloride (NaCl) can be added to increase in the number of ions in solution. But this leads to oxidation of chloride ions to chlorine gas and OCl ions, which are strong oxidants capable of oxidizing organic molecules present in the effluent [54].

The reactions (2, 3 and 4) are [29]:

$$Cl^-_{(aq)} \rightarrow Cl_{2(g)} + 2e^- \tag{2}$$

$$Cl_{2(g)} + H_2O \rightarrow HOCl_{(l)} + H^+_{aq} + Cl^-_{(aq)} \tag{3}$$

$$HOCl_{(l)} \rightarrow H^+_{aq} + OCl^-_{(aq)} \tag{4}$$

According to [55], the power consumption does not diminish significantly when the conductivity of the solution is greater than 1.5

mS/cm.

The conductivity of the effluent, namely the capacity to conduct electrical current, should be directly proportional to the amount of ions present in the conductive liquid. These ions are responsible for conducting the electrical current. It is evident, then, that the higher the concentration of these ions in the effluent, the greater its ability to conduct electrical current and the greater the possibility of reactions between the substances present in the effluent, a positive factor which enables reduction of energy consumption.

Temperature

According to [56], the effect of temperature has as yet been little investigated in the electrocoagulation process. Some studies have shown that the efficiency achieved with aluminum electrodes increases with temperature up to the 60 °C, above which the efficiency decreases. However, the conductivity increases with increasing temperature, decreasing the resistivity and electric power consumption. Increasing the temperature of the solution contributes to increase the efficiency of removal, caused by the increase of the movement of the ions produced, facilitating their collision with the coagulant formed [57, 58].

Electrolytic Processes Applied to the Treatment of Oily Wastewater

According to [36] in the 1980s, Zhdanov used iron and aluminum electrodes to break down emulsions and promote flocculation of wastewater impurities from drilling platforms, aiming at its reuse.

The EPA (1993) conducted studies on the use of innovative technologies for treating hazardous waste, using the technique of electrocoagulation with alternating current. The resulting apparatus was called the ACE Separator™. This technology introduces low concentrations of nontoxic aluminum hydroxide in the medium.

The effluents were prepared in order to reproduce the natural leakage to the underground reservoirs in soil washing operations. The main objective of these tests was to obtain optimal conditions for breaking oil-in-water emulsions and achieve reductions of soluble solids and loads of metal pollutants.

Experiments were carried out using a monopolar electrode of aluminum and the effluent used was prepared with 1.5% of diesel oil, 0.1% surfactant, 10 to 100 mg L of metals (Cu, Cd and Cr) and 3% soil containing 50% clay. Assays were performed at pH 5, 7 and 9. NaCl was added in the range from 1200 to 1500 mg/L to simulate salinity values found in contaminated media. The optimum operational conditions were: 4 A (ampere); space between the electrodes of 0.5 cm, duration of 3 to 5 minutes and frequency of 10 Hz. The pollutant removal efficiencies were 98% of TSS, 95% for TOC, 72% for Cu, 92% for Cr and 70% for Cd. However, fouling was observed on the plates of the electrodes.

Another research team [36] used the electrolytic process with titanium electrodes to promote the oxidation of pollutants in the oil industry. The process was tested with effluents of low and high salinity containing sulfides, ammonia and phenol, besides organic matter. Studies were also performed simulating the generation of chlorine by electrolysis with salinity levels similar to those found in the effluent.

The results demonstrated the possibility of using the electrolyte process in both situations, but that it was particularly advantageous when used in high salinity effluent due to the high conductivity, which allows oxidation with lower energy consumption.

[39] conducted experiments with electrolysis to remove the COD, O&G and turbidity from olive oil residues in the presence of H_2O_2 and a flocculating agent generated "in situ" via iron and aluminum electrodes. The iron electrode was more effective than aluminum. The COD removal efficiency was around 62-86%, whereas the removal of O&G, and turbidity was 100%. The current density ranged from 20-75 mA cm^2, depending on the concentration of H_2O_2 and coagulating agents. Using petrochemical industry effluents, [59] conducted tests of chemical coagulation (jar test) and electrocoagulation on a laboratory scale. The tests allowed comparing the removal efficiencies of organic matter by electrocoagulation and chemical coagulation, and comparing the efficiencies of these treatments in laboratory scale with those obtained in the stage of physical-chemical treatment (chemical coagulation and flocculation). In all cases the efficiency of removal of organic matter were evaluated according to reduction of COD.

In the chemical coagulation assays, the authors used aluminum sulfate. The parameters evaluated were the optimum pH for coagulation and optimum coagulant dosage. Tests were carried out of the electrolytic process in batch with aluminum electrodes. The parameters analyzed were: temperature, applied potential, initial pH, distance between electrodes, number of electrodes and electrode wear. The efficiency of the electrocoagulation process showed values up to three times higher than the monthly average obtained by the petrochemical industry using chemical coagulation and flocculation.

Another research team [13] studied the possibility of applying electrocoagulation to treat synthetic wastewater in the oil industry. This effluent was prepared in the proportion of 33 L of water to 50 ml of crude oil in a vessel with mechanical stirring for 30 minutes. A single-compartment electrochemical cell was used to generate bubbles, operating in continuous system with power feed at the top and treated effluent outlet at the bottom. The anode was made from titanium, called a DSA® and the cathode material was grade 316 stainless steel.

Electrolysis was carried out using current density of 20 mA/cm², flow rates of 800 and 1200 mL/h^{-1} and electrolysis times 150 and 180 minutes, respectively. The results showed that it was possible to obtain removal of COD and O&G greater than 90%.

Other tests have been performed to investigate the treatment of oily wastewater from washing the holds of ships, using the technique of electrocoagulation. The process was evaluated in laboratory scale and involved the use of two types of electrodes (iron and aluminum). The results showed that the best performance was obtained using the iron electrode [60].

The system operated with current of up to 1.5 A, for 60 and 90 minutes. The removal rates of BOD and O&G were 93 and 96%, respectively, while the COD removal rates were 61 and 78%, depending on the treatment time. Finally, 99% of the hydrocarbons were removed. Electrocoagulation was also effective for clarification of the effluent. Removal rates of 99 and 98% were measured for TSS and turbidity, respectively.

To verify the efficiency of treatment by synthetic cutting fluid, electroflotation was performed to characterize the fluid before and after treatment. The parameters analyzed were: pH, turbidity, metals, total phosphorus, COD, BOD and O&G. The results were quite satisfactory. The EF efficiency showed partial removal of contaminants in the cutting fluid, but the concentration of O&G exceeded the maximum limit for disposal according to relevant legislation [61].

Using the EC technique with a perforated aluminum electrode to separate oil emulsified in water, [62] found that the perforated electrode facilitates the passage and upward movement of the oil droplets to the surface. The authors observed that at 5 V and 0.4 A, the oil removal efficiency was 90% at pH 4.7 during 30 min of electrolysis, and the optimal salt concentration was 4 mg / L. The oil removal rate from the effluent increased with decreasing pH and lower salinity.

With the goal of removing Cu^{+2}, Zn^{+2}, phenol and BTEX from the produced water, [63] studied two types of electrochemical reactors: one using electro flocculation and the other electro flotation. In the former, an electric potential was applied to a solution containing NaCl, through electrodes of Fe, which with the dissolution of the metal ions generated Fe^{+2} and gases. An appropriate pH, these gases caused coagulation/flocculation reactions, removing Cu^{+2} and Zn^{+2}. In the second setup, a carbon steel cathode and DSA®. of DSA®Ti/TiO_2-RuO_2-SnO_2 were used, in a solution containing NaCl, which produced strong oxidizers such as HOCl and Cl_2. These promoted degradation of BTEX and phenol at different flow rates. The Zn^{+2} was removed by electrode position or by the formation of $Zn(OH)_2$ due to the increased pH.

Assessing the removal of oil from a synthetic emulsion by the electrocoagulation-flotation process, [45] observed the influence of operating parameters on the rate of reduction of COD, initial oil concentration, current density, electrode separation, pH and electrolyte concentration. NaCl was added to increase the solution's conductivity. The initial pH of the emulsion was 8.7. The Zeta potential had an average value of -75 mV, indicating emulsion stability. The author found that the best conditions for removal were current density of 4.44 mA/cm^2 treatment time of 75 min, distance between electrodes of 10 mm and concentration of the electrolyte (NaCl) of 3 g/L.

Studying the treatment of a synthetic effluent and a real produced water sample for removal of oil by the Fenton process, electro flotation and a combination these two, [64] first evaluated the Fenton and electro flotation processes individually and optimized the parameters for evaluating the combined process. The Fenton process, using Fe^{+2} and H_2O_2, obtained a peak oil removal of around 95% after 150 minutes and 50% removal after 57 minutes. The EC with the optimized volt (V) value managed to remove 98% of the oil after 40 minutes. The combined process using the optimized parameters for each process achieved removal of 98% after 10 minutes and 50% after 1 to 3 minutes. The combined process proved to be much more efficient than the procedures alone.

[65] evaluated the removal of sulfate and COD from oil refinery wastewater through three types of electrodes: aluminum, stainless steel and iron. They investigated the effects of current density, electrode array, electrolysis time, initial pH and temperature for two samples of wastewater with different concentrations of COD and sulfate. The experimental results showed that the aluminum anode and cathode was more efficient in the reduction of both contaminants. The results demonstrated the technical feasibility of electrocoagulation as a reliable method for pre-treatment of contaminated wastewater from refineries.

[66] in their experiments showed that treatment of synthetic wastewater emulsified water produced by EF, produced better results when used at a frequency of 60 Hz alternating current, initial pH 9, electrolysis time of 3 minutes and application of intensity current of 3 A. The results of tests on simulated wastewater produced water resulted in high removal efficiencies of organic load reaching 99% removal of oil and grease, color and turbidity. Compared to the flocculation trials using *Jar-Test*, the EF demonstrated highly efficient for the treatment of effluent water production in order to remove oil and grease emulsions, color and turbidity with no addition of chemical reagents or pH adjustment. *Jar-Test* trials were not effective consume high amounts of aluminum sulfate and low efficiency of removal of parameters. The main advantage of alternating current electrolysis in comparison with the direct current is less wear of the electrode mass. By using the same assay conditions for both technologies in 60 minutes oxidized alternating current of 1.6 g Al electrode while the oxidized direct current electrode of 3.4 g Al.

VARIABLE FREQUENCY AC ELECTRO FLOCCULATION

In this work, we used variable frequency electroflocculation, which consists of using alternating current from the power grid at 60 Hertz and varying the voltage and frequency between 1 Hz to 120 Hz. This alternating current was generated by reconstituting the sinusoidal form of the input current in a conversion system with vector control, which generates a pulse-controlled formation time (period) adjusted by a programmable base time through a system of microprocessors.

This system triggers an oscillator to form a new waveform that has a peak residence time large enough to have conduction at a given polarity. The evaluation of EF with alternating current for the treatment of effluents from oil platforms can be of great importance to develop treatment processes that are fast, efficient and cost-effective. The aim of this experiment was to develop and evaluate in laboratory scale a variable frequency electroflocculation (EF) system for the treatment of oily water generated in offshore oil production and to compare the results against those produced using direct current.

Electroflocculation Units

AC and DC electroflocculation units consisted of a glass electrolytic cell with capacity of 1 liter under magnetic stirring, in which an electrode was inserted vertically (monopolar in parallel) in a honeycomb arrangement, made of seven interspersed aluminum plates. These plates measured 10 cm long, 5 cm wide and 3 mm thick and were separated by spacers of 1 cm each. After a predetermined electrolysis period, we waited for 30 minutes for complete flotation of the emulsion to occur. Through a tap at the bottom of the beaker, the treated effluent was removed to assess the efficiency of electro flocculation (EF), which was done by monitoring, in triplicate, the following parameters: pH, conductivity, turbidity and color.

DC Unit

The DC electro flocculation unit used a voltage of up to 15 V. First AC power (110/220 V) was applied to a potentiometer connected to a step-down transformer, feeding the secondary stage rectifier bridge responsible for providing DC power to the electrodes by a polarity reversing switch, connected to a meter showing voltage (V) and current (A). These readings guide the operator regarding the parameters of honeycomb electrode array. Figure 4 shows the diagram of the experimental DC setup.

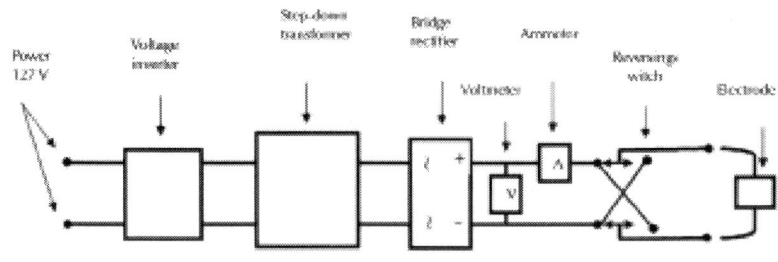

Figure 4: Schematic diagram of the experimental DC unit.

Variable Frequency AC Unit

The alternating current at a voltage of up to 15 V and variable frequency between 1 Hz and 120 Hz was obtained from a Weg CFW0800 AC/AC converter and a step-down transformer (Tecnopeltron PLTN model 100/15). In this setup, the input power at 60 Hz from the grid is converted to variable frequency output of 1 to120 Hz in order to obtain AC power at the desired level. As with the DC setup, there is a meter to indicate the voltage (V) and current (a), to guide the operator.

Figure 5 shows a block diagram where the 60 Hz current from the grid feeds a frequency converter with variable output from 1 Hz to 120 Hz, connected to a variable voltage step-down transformer, thereby providing appropriate frequency and voltage to the electrode. In the rectification step that occurs in the variable frequency converter, the power is transformed into DC. Then the new direct current is treated in the oscillator module which converts it into pulses with controlled width, forming a new AC waveform, with a frequency that can vary between 1 Hz and 120 Hz depending on the level of feedback (reference) from the load controller. Thus, it has a sinusoidal waveform where the period varies with the load, to obtain the best performance at active power levels.

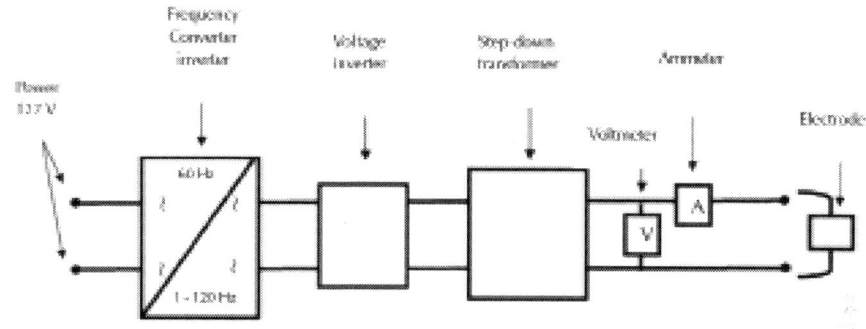

Figure 5: Schematic diagram of the experimental AC unit

The electrode is the central element for treatment. Thus, the proper selection of its materials is very important. The most common electrode materials for electro flocculation are aluminum and iron, since they are inexpensive, readily available and highly effective. In this experiment we used a hive array of seven interspersed aluminum plates measuring 10 cm long, 5 cm wide and 3 mm thick. The plates were separated by spacers (1 cm thick each), allowing varying the distance between the electrodes.

The electrodes were connected to specific instruments to control and monitor the current and voltage applied to the system, namely a frequency converter/regulator, potentiometer, step-down transformer, voltmeter, ammeter, bridge rectifier and polarity reversing switch.

Figure 6 shows an example of hive aluminum electrodes.

Figure 6: Hive aluminum electrodes containing eigth plates.

Tests with Real Effluent Using Oily Water from the Oil Industry

Through laboratory tests we noted that the real effluent yielded obtained from an oil company did not have high salinity and had high oil and grease content (60 g / L), so it was not characterized as produced water but rather as oily effluent.

We conducted there tests:

- the first using the effluent as received;
- the second adding 60g / L of natural salt; and
  1. the third adding salt 60g / L of salt plus emulsifiers.

Table 1 shows the values obtained with the AC and DC electroflocculation processes with the original effluent as received. The results of the AC setup were obtained with the maximum current of the unit (i = 2.5 A), due to the low salinity. The voltage was 11 V. In the case of direct current, the unit only reached a maximum of 1.6 A, so we added 1 g of salt to obtain the same current intensity as the AC unit. By adding salt to the effluent in the DC system, there is an improvement in removal efficiency.

The analysis of the oil and grease parameter (supernatant) was carried out separately, while the remaining parameters were analyzed with the subnatant phase of the effluent. The AC and DC electroflocculation tests were performed with the effluent containing an oil and grease content of 60g/L.

There was an increase in pH during the final AC and DC tests, attributed to the generation of OH-ions during the water reduction step.

$$2H_2O_{(l)} + 2e^- \rightarrow H_{2(g)} + 2OH^-_{(aq)}$$

(6)

Although these ions are also used to form the coagulating agent, the remaining quantity results in an increase of the pH value. This was also observed by [43, 67, 68].

Table 1: Data from the EF treatment process

Parameters	Oily wastewater	AC	DC
pH	6.7	8.3	8.2
Turbidity (NTU)	840	2	2
Color $_{(Abs. 400 nm)}$	0.46	0.02	0.01
Salinity (mg/L)	279	340	1210
Conductivity (μS/cm)	580	702	2238
TDS (mg/L)	408	498	1680
Phenols (mg/L)	0.5	< 0.1	< 0.1
Sulfides (mg/L)	2.8	< 0.5	< 0.5
Ammonia nitrogen (mg/L)	36.0	3.7	1.5
O&G (mg/L)	60000	18	22
Current (A)	0	2.5	2.5
Tension (V)	0	11.0	8.5

Legend: Data: time = 15 min., Delectrode. = 1 cm, vol. = 3L, i = 3 A, f = 60 Hz AC. Note: In the DC setup, 1 g of salt was added to increase the conductivity and the current intensity of the equipment, because it was unable to reach the same as with the AC setup.

Despite the low conductivity of the effluent, there was high removal of pollutants and complete clarification after treatment for 15 minutes. The removal of ammonia, phenols and sulfides may have been obtained by drag of the gas phase (electro flotation).

Figure 7 shows the evolution of the tests of the raw effluent and after treatment with AC and DC.

Figure 7: EF tests with real effluent with low salinity. Legend: a) raw effluent; b) effluent treated with AC electroflocculation; c) effluent treated with DC electroflocculation; and d) comparison of raw effluent with samples treated with AC and DC electroflocculation. Note: The tests performed with t = 15 min.

The test results using the effluent plus NaCl to bring the salinity to 60 g / L, to simulate produced water, are shown in Table 2.

Table 2: Results obtained during tests of AC and DC electro flocculation using effluent with high salinity.

Parâmeters	Oily wastewater	AC	DC
pH	5.9	6.3	6.3
Turbidity (NTU)	1000	10	16
Color (Abs.)	0.63	0.03	0.05
Salinity (g/L)	50.8	46.0	47.2
Conductivity (mS/cm)	91.5	72.4	75.3
TDS (g/L)	64.9	59.6	59.6
Phenols (mg/L)	0.5	< 0.1	< 0.1
Sulfide (mg/L)	2.8	< 0.5	< 0.5
Ammonia nitrogen (mg/L)	36.0	2.9	2.7
O&G (mg/L)	60000	30	35
Mass electrode (g)	-	0.16	0.23
Current (A)	-	2.5	2.5
Tension (V)	-	1.5	2.0

Data: vol. = 3 L, t = 6 min., Delectrode. = 1 cm, vol. = 3 L, i = 3 A, f = 60 Hz AC.

Since the electrolytic process involves corrosion of the electrode, according to Faraday's laws, there is mass loss of the electrodes. We measured this electrode mass loss by the weight difference before and after each test. High removal efficiencies were observed in tests with the high-salinity effluent. Electrode mass consumption with alternating current was 33% lower than with direct current, with all other conditions the same. Low voltage was applied during electrolysis in the tests to achieve high pollutant removal efficiency. In previous tests, the low conductivity greatly increased the voltage required by the system, leading to high energy consumption. This indicates that high conductivity greatly favors the electrolytic process.

Figure 8 shows the evolution of the tests with high-salinity effluent using the EF technology with alternating current and direct current. Note the total clarification after the tests without filtration.

Table 3 shows the results of the high-salinity effluent treated with the AC and DC processes.

As can be seen from the above table, the high turbidity, color and O&G were almost completely removed (above 99%). The voltage applied to the electrodes was very low, resulting in high efficiency of the technique. The electrode mass consumed with DC was 31% higher than with AC.

A hypothesis for the lower electrode consumption with alternating current is that since DC only flows in one direction, there may be irregular wear on the plates due to the onslaught of the current and subsequent oxidation occurring in the same preferential points of the electrode. In the case of AC, the cyclical energization retards the normal mechanisms of attack on an electrode and makes this attack more uniform, thus ensures longer electrode life.

Figure 8: EF tests with high-salinity effluent. Legend: 60g / L NaCl: a) raw wastewater containing 60g / L of O&G; b) raw wastewater being mixed salt, for EF testing; c) effluent treated with AC (left) and DC (right) electro flocculation.

Table 3: Results of treating high-salinity effluent with AC and DC electrolysis

Parâmeters	Oily wastewater	AC	DC
pH	6.4	6.7	6.7
Turbidity (NTU)	11050	8	10
Color (Abs.)	7.57	0.03	0.04
Salinity (g/L)	50.8	47.5	47.8
Conductivity (mS/cm)	91.5	87.4	87.0
TDS (g/L)	64.9	61.9	61.5
Phenols (mg/L)	0.5	0.2	0.2
Sulfide (H_2S) (mg/L)	2.8	1.2	1.0
Ammonia nitrogen (mg/L)	36	4	5
O&G (mg/L)	60000	32	30
Mass electrode (g)	-	0.18	0.26
Current (A)	-	2.5	2.5
Tension (V)	-	1.5	2.0

Note: Data: vol. = 3 L, t = 6 min

Figure 9 shows the evolution of EF treatment of high salinity effluent during the stages of development using AC and DC.

Figure 9: Tests of EF with high-salinity effluent. Legend: (60g/L) of salt. Processing steps: a) emulsified effluent, b) effluent undergoing electro floccula-

tion with formation of supernatant sludge, c) treated effluent, with sludge formation; d) original effluent and after treatment with AC (left) and DC (right) electro flocculation.

CONCLUSIONS

In the present study, we could confirm that the EF process produces satisfactory results for treatment of oily wastewater, allowing its discharge into water bodies or reinjection in oil formations. The AC technology was highly effective, both with the original oily water as received and with the simulated produced water after addition of salt.

Overall, the results confirm the potential of the technique, which through simple and compact equipment, can be employed for the decontamination of organic compounds. The results of tests on oily water resulted in high organic load removal efficiencies, reaching 99% removal of oil and grease, color and turbidity, along with high removal of phenols, ammonia and sulfides.

The biggest advantage of AC versus DC electroflocculation is the lower electrode wear with the former technique. When using the same testing conditions and time of 6 minutes for both technologies, the efficiency was above 30%. The AC electroflocculation technique seems to be a promising alternative in the treatment of oily wastewater from the oil industry.

ACKNOWLEDGEMENTS

The National Councel of Technological and Scientific Development (CNPQ) When National Institute of Oil and Grease (INOG), we thank the Environmental Technology Laboratory (LABTAM) for the tests performed during this study, the Graduate Program in Environmental Protection (PPG-MA/UERJ) for support, and Rio de Janeiro State Research Foundation (FAPERJ) for financing through a PhD scholarship.

REFERENCES

1. Limons L S2008Avaliação do potencial de utilização da macrófita aquática seca salvinia sp. No tratamento de efluentes de fecularia. 87f. Dissertação (Mestrado)- Universidade Estadual do Oeste do Paraná, Cascavel, Brazil.

2. Mariano J B2005Impactos ambientais do refino de petróleo. Rio de Janeiro: Interciência.

3. Magossi L R, Bonacella P H1999a poluição das águas. 17 ed. São Paulo: Ed. Moderna. (Coleção Desafios)

4. J. Barboza, Impactos ambientais do refino de Petróleo. Rio de Janeiro: Interciência, 2005

5. A. L. O. Campos, et al.2005Produção mais limpa na industria de petróleo: o caso da água produzida no campo de Carmópolis/SE. In: Congresso Brasileiro de Engenharia Sanitária e Ambiental, 23, Campo Grande, MS. Anais, ABES.

6. Thomas J E2004Fundamentos de Engenharia de Petróleo. 2. ed. Rio de Janeiro: Interciência.

7. P. Cañizares, et al.2007Break-up of oil-in-water emulsions by electrochemical techniques. J. Haz. Materials, 145n. 1/2, 233240

8. A. Pinotti, N. Zaritzky, 2001Effect of aluminium sulphate and cationic polyelectrolytes on the destabilization of emulsified waste. Waste Manage, 21535542

9. G. Rios, C. Pazos, J. Coca, 1998Electrolytic Treatment of Wastewater in the Oil IndustryColloid Surf. v. A 138, 383389

10. M. E. T. A. L. Carmona, 2006Electrolytic Treatment of Wastewater in the Oil IndustryChem. Eng. Science, 6112331242

11. L. S. Calvo, et al.2003Electrolytic Treatment of Wastewater in the Oil IndustryEnvironmental Program. 225765

12. B. R. Kim, et al.1992Electrolytic Treatment of Wastewater in the Oil Industryr. W. Env. Research, 64216222

13. A. C. Santos, et al.2007Tratamento de efluentes sintéticos da indústria do petróleo utilizando o método de eletroflotação. In: PDPETRO, 4, 2007, Campinas. Anais. Campinas, ABPG.

14. M. Y. A. Mollah, et al.2001Electrolytic Treatment of Wastewater in the Oil IndustryJ. Haz. Materials, 842941

15. Rosa A J2006Carvalho, R. S.; Xavier, J. A. D. Engenharia de reservatórios de petróleo. Rio de Janeiro: Interciência.

16. Macedo V A P2009Tratamento de água de produção de petróleo através de membranas e processos oxidativos avançados. 92f. Dissertação (Mestrado)- Universidade de São Paulo, Lorena, Brazil.

17. E. Oliveira, R. Santelli, R. Casella, 2005Direct determination of lead in produced waters from petroleum exploration by electrothermal atomic absorption spectrometry X-ray fluorence using Ir-W permanent modifier combined with hydrofluoric acid. Analyt. Chim. Acta, 5458591

18. Ramalho J B V S1992Curso básico de processamento de petróleo: tratamento de água oleosa. Rio de Janeiro: RPSE:DIROL:SEPET.

19. Gabardo I T2007Tese de Doutorado, Universidade Federal do Rio Grande do Norte, Brazil.

20. S. B. Henderson, et al.1999Potencial impact of production chemicals on the toxicity of produced water discharges from North Sea Oil Platforms. M. Pollut. Bulletin, 38n. 12, 11411151

21. H. Dórea, et al.2007Electrolytic Treatment of Wastewater in the Oil Industryl. Microchemical Journal85234238

22. J. C. Campos, et al.2002Electrolytic Treatment of Wastewater in the Oil IndustryWater Research3695104

23. Vieira D S, Cammarota M C, Camporese E F S2003Redução de contaminantes presentes na água de produção de petróleo. In: Congresso Brasileiro de P&D em Petróleo e Gás, 2, Rio de Janeiro. Anais. Rio de Janeiro: Instituto Brasileiro de Petróleo e Gás.

24. G. M. A. Cunha, et al.2005Tratamento de águas produzidas em campos de petróleo: estudo de caso da estação de Guamaré/RN. In: Congresso Brasileiro de Engenharia Química em Iniciação Científica, 6, Campinas. Anais. Campinas, Unicamp.

25. P. Holt, 2002Electrolytic Treatment of Wastewater in the Oil Industryf. Thesis (Ph. D.)- University of Sydney, Austrália.

26. M. Y. A. Mollah, et al.2004Electrolytic Treatment of Wastewater in the Oil IndustryJ. Haz. Materials, 114199210

27. Wiendl W G1998Processos eletrolíticos no tratamento de esgotos sanitários. Rio de Janeiro: ABES.

28. M. L. Torem, et al.2002Remoção de metais tóxicos e pesados por eletroflotação. Saneamento Ambiental, 854651

29. J. Ge, J. Qu, P. Lei, J. Liu, 2004Electrolytic Treatment of Wastewater in the Oil Industryr. Sep. Purif. Technology, 363339

30. G. Chen, X. Chen, P. L. Yue, 2000Electrolytic Treatment of Wastewater in the Oil IndustryJournal Environmental Engineering. 126n. 9, 858863

31. P. R. Kumar, S. Chaudhari, K. C. Khilar, S. P. Mahajan, 2004Chemosphere, 9, 55.

32. Cerqueira A A, Russo C, Marques M R C2009Electrolytic Treatment of Wastewater in the Oil IndustryBraz. J. Chem. Engineering, 26659668

33. M. Murugananthan, G. Bhaskar, S. Prabhakar, 2004Electrolytic Treatment of Wastewater in the Oil IndustryJ. Haz. Materials, 1093744

34. C. Y. Hu, et al.2005Electrolytic Treatment of Wastewater in the Oil IndustryW. Research, 39n. 5, 895901

35. A. E. Yilmaz, R. Boncukcuoglu, M. M. Kocakerin, B. . Keskinler, 2005J. Haz. Mater., 125, 1.

36. Queiroz M S, Souza A D, Abreu E S V, Gomes N T, Neto O A A1996Aplicação do Processo Eletrolítico ao Tratamento de Água de Produção, CENPES-DITER-SEBIO, RT: Rio de Janeiro, Brazil.

37. S. Rubach, I. F. Saur, 1997Electrolytic Treatment of Wastewater in the Oil IndustryFilt. Separation, 34n. 8.

38. M. Khemis, et al.2005Electrolytic Treatment of Wastewater in the Oil Industrysuspensions. Trans IChemE, 83n. 81, 5057

39. Ü. T. Ün, et al.2006Electrolytic Treatment of Wastewater in the Oil IndustrySep. Purif. Technology, 52136141

40. K. Bensadok, et al.2008Electrocoagulation of cutting oil emulsion using aluminum plate electrodes. J. Haz. Materials, 152n. 1, 423430

41. Cerqueira A A, Marques M R C, Russo C2011Avaliação do processo eletrolítico em corrente alternada no tratamento de água de produção. Quimica Nova, 341

42. Silva A L C2002Processo eletrolítico: uma alternativa para o tratamento de águas residuárias. 60f. Monografia (Especialização

em Química Ambiental)- Universidade do Estado do Rio de Janeiro, Rio de Janeiro, Brazil.

43. M. Kobya, et al.2006Treatment of potato chips manufacturing wastewater by electrocoagulation. Desalination, 190n. 1-3, 201211

44. A. Dimoglio, et al.2004Petrochemical wastewater treatment by means of clean electrochemical technologies. Clean Tech. Envir. Policy, 6n. 4, 288295

45. Merma A G2008Eletrocoagulação aplicada aos meios aquosos contendo óleo. 128f. Dissertação (Mestrado)- Pontifícia Universidade Católica do Rio de Janeiro, Rio de Janeiro, Brazil.

46. Y. Avsar, U. Kurt, Gonullu, T(2007Electrolytic Treatment of Wastewater in the Oil IndustryJ. Haz. Materials. 148n.1/2, 340345

47. Rangel R M2008Modelamento da eletrocoagulação aplicada ao tratamento de águas oleosas provenientes das indústrias extrativas. 154f. Tese (Doutorado)- Pontifícia Universidade Católica do Rio de Janeiro, Rio de Janeiro, Brazil.

48. R. D. Letterman, A. Amirtharaj, Mélia. C. R. O`, 1999Coagulation and flocculation. In: Electrolytic Treatment of Wastewater in the Oil IndustryNew York: McHill.

49. Crespilho F N, Rezende M O O2004Eletroflotação: princípios e aplicações. São Carlos: Rima.

50. W. Den, C. Huang, 2005Electrocoagulation for removal of silica nano-particles from chemical-mechanical-planarization wastewater. Coll. Surfaces A: Physic. Eng. Aspects, 254n. 1/3, 8189

51. N. Modirshahla, M. A. Behnajady, S. Kooshaiian, 2007Investigation of the effect of different electrode connections on the removal efficiency of Tartrazine from aqueous solutions by electrocoagulation. Dyes and Pigments, 74n. 1/ 2, 249257

52. N. Daneshvar, A. Oladegaragoze, N. Djafarzadeh, 2006Electrolytic Treatment of Wastewater in the Oil IndustryJ. Haz. Materials, 129116122

53. N. Daneshvar, H. A. Sorkhabi, M. B. Kasiri, 2004Decolorization of dye solution containing Acid Red 14 by electrocoagulation with

a comparative investigation of different electrode connections. J. Haz. Materials, 112n. 1/2, 5562

54. A. K. Golder, et al.2005Electrolytic Treatment of Wastewater in the Oil IndustryJ. Haz. Materials, 127n. 1/3, 134140

55. P. Gao, et al.2005Removal of chromium (VI) from wastewater by combined electrocoagulation-electroflotation without a filter. Sep. Purif Technology, 43n. 2, 117123

56. G. Chen, 2004Electrolytic Treatment of Wastewater in the Oil IndustrySep. Purif. Technology, 381141

57. M. Y. Ibrahim, et al.2001Electrolytic Treatment of Wastewater in the Oil IndustrySep. Science Technology, 3637493762

58. N. Aneshvar, et al.2007Decolorization of C.I. Acid Yellow 23 solution by electrocoagulation process: investigation of operational parametersand evaluation of specific energy consumption (SEEC). J. Haz. Materials, 148566572

59. Wimmer A C S2007Aplicação do processo eletrolítico no tratamento de efluentes de uma indústria petroquímica. 2007. 195f. Dissertação (Mestrado)- Pontifícia Universidade Católica do Rio de Janeiro, Rio de Janeiro, Brazil.

60. M. Asselin, et al.2008Electrolytic Treatment of Wastewater in the Oil IndustryJ. Haz. Materials, n. 151, 446455

61. Fogo F C2008Avaliação e critérios de eficiência nos processos de tratamento de fluido de corte por eletroflotação. 103f. Dissertação (Mestrado)- Universidade de São Paulo, São Paulo, Brazil.

62. R. M. Bande, et al.2008Electrolytic Treatment of Wastewater in the Oil IndustryChem. Eng. Journal. 137n. 3, 503509

63. Ramalho A M Z2008Estudo de reatores eletroquímicos para remoção de Cu2+, Zn2=, Fenol e BTEX em água produzida. 86f. Dissertação (Mestrado)- Universidade Federal do Rio Grande do Norte, Natal, Brazil.

64. Gomes E A2007Tratamento combinado de água produzida de petróleo por electroflocculation e processo fenton. 84f. Dissertação (Mestrado)- Universidade Tiradentes, Aracaju, Brazil.

65. M. H. El -Naas, et al.2009Electrolytic Treatment of Wastewater in the Oil IndustryJ. Envir. Management, 91180185

66. Cerqueira A A2011Aplicação da técnica de eletrofloculação utilizando corrente alternada de frequência variável no

tratamento de água de produção da indústria do petróleo. 133f. Tese de Doutorado- Universidade do Estado do Rio de Janeiro, Rio de Janeiro, Brazil.

67.   Ferreira L H2006Remoção de sólidos em suspensão de efluente da indústria de papel por eletroflotação. 82f. Dissertação (Mestrado)- Universidade Estadual de Campinas, São Paulo, 2006.

68.   S. Irdemes, et al.2006The effect of current density and phosphate concentration on phosphate removal from wastewater by electrocoagulation using aluminum and iron plate electrode. Sep. Purif. Technology, 52

Chapter 5

Novel Formulation of Environmentally Friendly Oil Based Drilling Mud

Adesina Fadairo[1], Churchill Ako[1],
Abiodun Adeyemi[1], Anthony Ameloko[1],
and Olugbenga Falode[2]

[1]Department of Petroleum Engineering, Covenant University, Ota, Nigeria

[2]Department of Petroleum Engineering, University of Ibadan, Nigeria

INTRODUCTION

The term drilling fluids or drilling muds generally applies to fluids used to help maintain well control and remove drill cuttings (rock fragments from underground geological formations) from holes drilled in the earth. Drilling fluids are fluids used in petroleum drilling operations. These fluids are a mixture of clays, chemicals, water, oils. These fluids are used in a borehole during drilling operations for [1]:

- Hole cleaning
- Pressure control
- Cooling and lubrication of the bit
- Corrosion control (especially for oil-based muds)

- Formation damage control
- Wellbore stability maintenance
- Transmission of hydraulic energy to BHA (Bottom Hole Assembly)
- Aid in cementing operations
- Minimize environmental impact
- Inhibit gas hydrate formation in the well.
- Avoid loss of circulation and seal permeable formations.

Considering each of the uses, the primary use of drilling fluids is to conduct rock cuttings within the well. If these cuttings are not transported up the annulus between the drillstring and wellbore efficiently, the drill string will become stuck in the wellbore. The mud must be designed such that it can, carry the cuttings to surface while circulating, suspend the cuttings while not circulating, and drop the cuttings out of suspension at surface [1-5].

The hydrostatic pressure exerted by the mud column must be high enough to prevent an influx of formation fluids into the wellbore, but the pressure should not be too high, as it may fracture the formation. The instability caused by the pressure differential between the borehole and the pore pressure can be overcome by increasing the mud weight. The hydration of the clays can only be overcome by using non water-based muds, or partially addressed by treating the mud with chemicals which will reduce the ability of the water in the mud to hydrate the clays in the formation. These muds are known as inhibited muds. While drilling, the rock cutting procedure generates a lot of heat which can cause the bits, and the entire BHA (Bottom Hole Assembly) wear out and fail, and the drilling muds help in cooling and lubricating the BHA. These fluids also help in powering the bottom hole tools. In cementing operations, drilling fluids are used to push and pump the cement slurry down the casing and up the annular space around the casing string in the hole.

The drilling fluid must be selected and or designed so that the physical and chemical properties of the fluid allow these functions to be fulfilled. However, when selecting the fluid, consideration must also be given to [5-6]:

- The environmental impact of using the fluid
- The cost of the fluid
- The impact of the fluid on production from the reservoir

CLASSIFICATION OF DRILLING FLUIDS

Drilling fluids are classified according to the continuous phase [1, 3]

- The WBM (Water Based Muds), with water as the continuous phase.
- The OBM (Oil Based Muds), with oil as their continuous phase.
- The Pneumatic fluids (with gases or gas-liquid mixtures as their continuous phase)

This chapter narrows our focus to oil based drilling fluids (OBM).

In general, OBM are drilling fluids which have oil as their dominant or continuous phase. A typical OBM has the following composition:

Clays and sand about 3%, Salt about 4%, Barite 9%, Water 30%, Oil 50-80%.

OBM have a whole lot of advantages over the conventional WBM. This is due to the various desirable rheological properties that oils exhibit. Since the 1930s, it has been recognized that better productivity is achieved by using oil rather than water as the drilling fluid. Since the oil is native to the formation it will not damage the pay zone by filtration to the same extent as would a foreign fluid such as water. We shall outline some of the desirable properties of oil based muds, which include [4]:

- Shale Stability: OBM are most suited for drilling shaly formations. Since oil is the continuous phase & water is dispersed in it, this case results in non-reactive interactions with shale beds.
- Penetration Rates: OBM usually allow for increased penetration rates.
- Temperature: OBM can be used to drill formations where BHT (Bottom Hole Temperatures) exceed water based mud tolerances. Sometimes up to over 1000 degrees rankine.
- Lubricity: OBM produce thin mud cakes, and the friction between the pipe and the well bore is minimized, thus reducing the pipe differential sticking. Especially suitable for highly deviated and horizontal wells.

- Ability to drill low pore pressured formations is accomplished, since the mud weight can be maintained at a weight less than that of water (as low as 7.5 ppg).

- Corrosion control: Corrosion of pipes is reduced since oil, being the external phase coats the pipe. This is due to the fact that oils are nonconductive, thermally stable, and more often, do not permit microbial growth.

- OBM can be re used, and can also be stored for a long period of time since microbial activity is suppressed.

The basic kind of oil used in formulating OBM is the diesel oil, which has been in existence for a long time, but over the years, diesel oil based muds have posed various environmental problems.

Water-based muds (WBMs) are usually the mud of choice in most drilling operation carried out in sandstone reservoir, however some unconventional drilling situations such as deeper wells, high temperature/pressure formation, deepwater reservoir, alternative shale-sand reservoir and shale resource reservoir require use of other mud systems such as oil based mud to provide acceptable drilling performance [5-8].

OBM is needed where WBM cannot be used especially in hot environment and salt beds where formation compositions can be dissolved in WBM. OBM have oil as their base and therefore more expensive and require more stringent pollution control measures than WBM.

It is imperative to propagate the use of environmentally friendly and biodegradable sources of oil to formulate our OBM, thereby making it less expensive and environmentally safe and equally carry out the basic functions of the drilling mud such as maintenance of hydrostatic pressure, removal of cuttings, cooling and lubricating the drill string and also to keep newly drilled borehole open until cementing is carried out.

Background

Environmental problems associated with complex drilling fluids in general, and oil-based mud (OBM) in particular, are among the major concerns of world communities. Among others are the problems faced by some host communities in the Niger Delta region of Nigeria. For this reason, the Environmental Protection Agency (EPA) and other regulatory bodies are imposing increasingly stringent regulations to ensure the use of environmentally friendly muds [7-8].

Throughout the 1970s and 1980s, the EPA and other regulatory bodies imposed environmental laws and regulations affecting all aspects of petroleum-related operations from exploration, production and refining to distribution. In particular, there has been increasing pressure on oil and gas industry stakeholders to find environmentally acceptable alternatives to OBMs. This has been reflected in the introduction of new legislation by government agencies in almost every part of the world.

The researches and surveys conducted came up with possibilities of having environmentally friendly oil based mud. Stakeholders in the oil and gas industry have been tasked with the challenge of finding a solution to this problem by formulating optimum drilling fluids and also reduce the handling costs and negative environmental effects of the conventional diesel oil based drilling fluid. An optimum drilling fluid is one which removes the rock cuttings from the bottom of the borehole and carries them to the surface, hold cuttings and weight materials in suspension when circulation is stopped (e.g during shut in), and also maintain pressure. An optimum drilling fluid also does this at minimum handling costs, bearing in mind the HSE (Health, Safety, Environment) policy in mind [6].

In response to the harmful effects of diesel oil on the environment and on the ozone layer (as a result of the emission of greenhouse gases), researches and surveys have gone on in the past two to three decades, and have come up with mud formulations based on the use of plant oils as diesel substitutes. Over the years, plant oils have become increasingly popular in the raw materials market for diesel substitutes. The most popular being: Rapeseed oil, Jatropha oil, Mahua oil, Cottonseed oil, Sesame oil, Soya bean oil, palm oil etc. This brings about the importance of agro allied intervention in the energy industry. Hence, the contribution of non-edible oils such as jatropha oil, canola oil, algae oil, moringa seed oil and Soapnut will be significant as a plant oil source for diesel substitute production.

This chapter describes the formulation of environmental friendly oil based mud (using plant oil such as jatropha oil, algae oil and moringa seed oil) that can carry out the same functions as diesel oil based drilling fluid and equally meet up with the HSE (Health, Safety and Environment) standards. Mud tests have been carried out at standard conditions on each plant oil sample so as to ascertain the rheological properties of the drilling fluid formulations. The conventional diesel oil based mud would serve as control.

Motivation

Drilling mud is in varying degrees of toxicity. It is difficult and expensive to dispose it in an environmentally friendly manner. Protection of the environment from pollutants has become a serious task. In most countries like Nigeria, the drilling fluids industries have had numerous restrictions placed on some materials they use and the methods of their disposal. Now, at the beginning of the 1990's, the restrictions are becoming more stringent and restraints are becoming worldwide issues. Products that have been particularly affected by restrictions are oil and oil-based mud. These fluids have been the mud of choice for many environments because of their better qualities. Initially, the toxicity of oil-based fluids was reduced by the replacement of diesel oil with low-aromatic mineral oils. In most countries today, oil-based mud may be used but not discharged in offshore or inland waters. Potential liability, latent cost, and negative publicity associated with an oil-mud spill are economic concerns. Consequently, there is an urgent need for the drilling fluids industry to provide alternatives to oil-based mud.

Methodology of the Study

Four different mud samples were mixed, and the base fluid was varied. The base fluids were algae, moringa, diesel and jathropha oils used in formulating the muds in an oil water ratio of 70:30, where diesel based mud served as the control.

The following equipment and materials were used to carry out the experiment:

Table 1: Materials and Apparatus required

Materials	Equipment
Pulverized bentonite	Weighing balance
Barite	Retort
Diesel oil	Halminton Beach Mixer
Canola oil	Condenser
Castor oil	Mud balance
Jatropha seeds	Round bottom flask
Water	Rotary viscometer
n-hexane	Resistivity meter
Filter paper	API filter press
Threads	pH meter
Universal pH paper strips	Soxhlet extractor
Algae	Heating mantle
	Vernier Caliper
	Reagent bottles

Experimental Procedure

The plant seeds (jatropha, moringa and algae) were collected from the western part of Nigeria, peeled and dried in an oven at about 55°C for seventy minutes. The dried seeds were then de-hulled, to remove the kernels. The brownish inner parts of the kernels were ground in a blender (to increase the surface area for the reaction).

Extraction

The method employed in this study is solvent extraction. Solvent extraction is a process which involves extracting oil from oil-bearing materials by treating it with a low boiling point solvent as opposed to extracting the oils by mechanical pressing methods (such as expellers, hydraulic presses, etc.). The solvent extraction method recovers almost all the oils and leaves behind only 0.5% to 0.7% residual oil in the raw material. Here the equipment used was the Soxhlet extractor. A Soxhlet extractor is a piece of laboratory apparatus invented in 1879 by Franz von Soxhlet. It was originally designed for the extraction of a lipid from a solid material.

Figure 1: Soxhlet extractor assembly.

The extraction procedure is given below:

- 50g of crushed plant seeds were measured out, and tied in filter papers.
- The sample was loaded into the main chamber of the Soxhlet extractor and poured in about 300ml of n-Hexane through the main chamber.
- The chamber is fitted into a flask containing 300ml of n-Hexane.
- The heating mantle was turned on and the system was left to heat at 70° C. The solvent was heated to reflux. The solvent vapour travelled up a distillation arm, and flooded into the chamber housing the solid wrapped in filter papers. The condenser condensed the solvent vapour, and the vapour dripped back down into the chamber housing the solid material.
- Then at a certain level, the siphon emptied the liquid into the flask.
- This cycle was repeated until the sample in the chamber changed colour to a considerable extent, and collected the fluid mixture in glass reagent bottles.
- The mixture was separated via the use of simple distillation, as shown in the set up in Fig. 2.
- The distillation took place at 70°C; the hexane was recovered and re-used while the oil was stored.

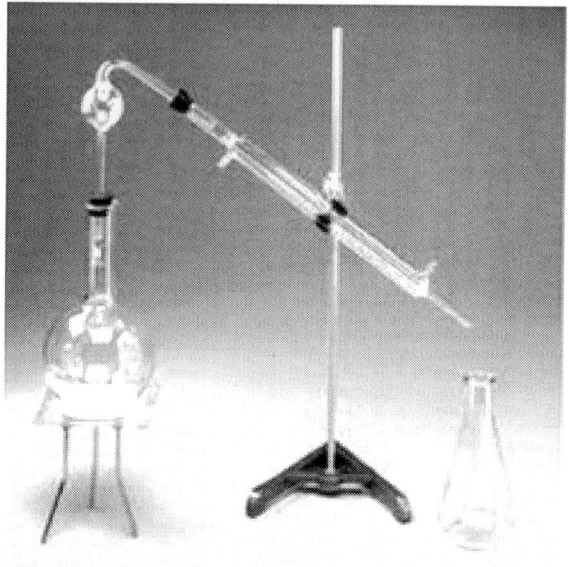

Figure 2: Set-up for distillation.

MUD PREPARATION

The densities of the various base fluids (water, algae oil, moringa oil, jatropha oil and diesel) were measured using the mud balance shown in diagram 3

- Using the weighing balance, the various quantities of materials as shown in Table 2 below were measured.
- The quantities of water and oil were measured using measuring beakers.
- Using the Hamilton Beach Mixer, the measured materials were thoroughly mixed until a homogenous mixture was obtained.
- The mud samples were aged for 24 hours.

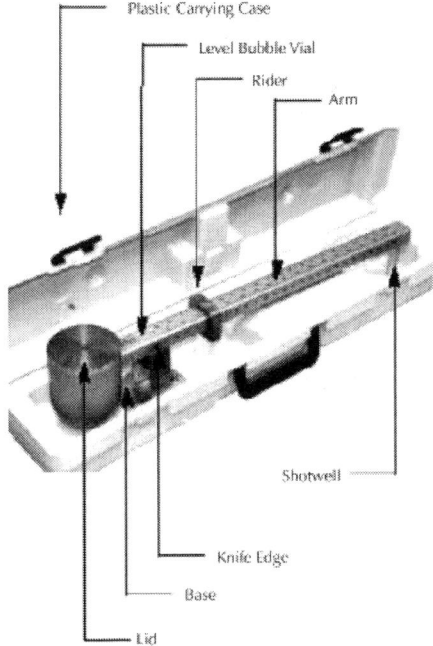

Plastic Carrying Case
Level Bubble Vial
Rider
Arm
Shotwell
Knife Edge
Base
Lid

Figure 3: Mud Balance

Density

- The aged mud samples were agitated for 2 minutes using the Hamilton Beach Mixer.
- The clean, dry mud balance cup was filled to the top with the newly agitated mud.
- The lid was placed on the cup and the balance was washed and wiped clean of overflowing mud while covering the hole in the lid.
- The balance was placed on a knife edge and the rider moved along the arm until the cup and arm were balanced as indicated by the bubble.
- The mud weight was read at the edge of the rider towards the mud cup as indicated by the arrow on the rider and was recorded.
- Steps 2 to 5 were repeated for the other samples.

Viscosity

- The mud was poured into the mud cup of the rotary viscometer shown in Diagram 4, and the rotor sleeve was immersed exactly to the fill line on the sleeve by raising the platform. The lock knot on the platform was tightened.
- The power switch located on the back panel of the viscometer was turned on.
- The speed selector knob was first rotated to the stir setting, to stir the mud for a few seconds, and it was rotated at 600RPM, waiting for the dial to reach a steady reading, the 600 RPM reading was recorded.
- The above process was repeated for 300 RPM, 200 RPM, 100 RPM, 60 RPM, 30 RPM and 6 RPM.
- Steps 7 to 10 were repeated for other samples.

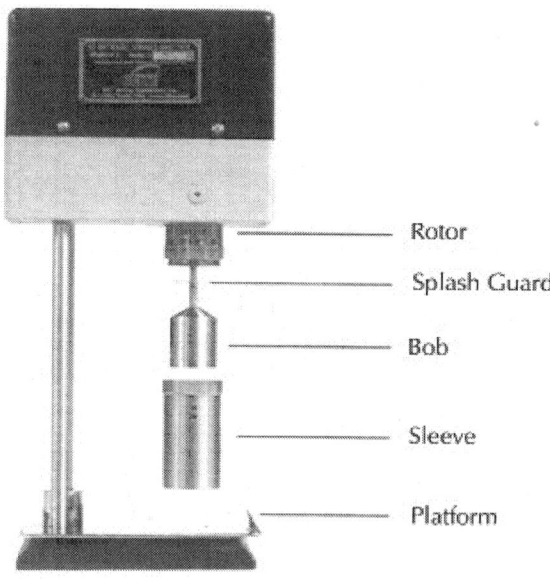

Figure 4: Rotational Viscometer

Gel Strength

The speed selector knob was then rotated to to stir the mud sample for a few seconds, then it was rotated to gel setting and the power was immediately shut off.

- As soon as the sleeve stopped rotating, the power was turned on after 10 seconds and 10 minutes respectively. The maximum dial was recorded for each case.
- Steps 12 and 13 were repeated for other samples.

Mud Filtration Properties

- The assembly is as shown in fig 5
- Each part of the cell was cleaned, dried and the rubber gaskets were checked.
- The cell was assembled as follows: base cap, rubber gasket, screen, filter paper, rubber gasket, and cell body.

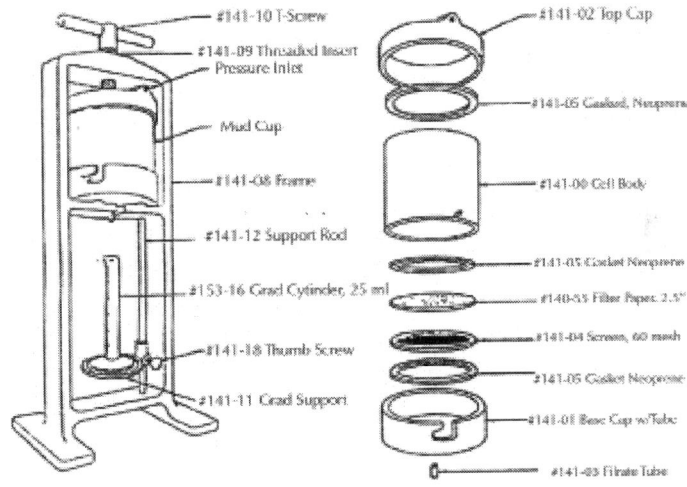

Figure 5: API Filter Press

- A freshly stirred sample of mud was poured into the cell to within 0.5 inch (13 millimeters) to the top in order to minimize contamination of the filtrate. The top cap was checked to ensure that the rubber gasket was in place and seated all the way around and complete the assembly. The cell assembly was placed into the frame and secured with the T-screw.

- A clean dry graduated glass cylinder was placed under the filtrate exit tube.
- The regulator T-screw was turned counter-clockwise until the screw was in the right position and the diaphragm pressure was relieved. The safety bleeder valve on the regulator was put in the closed position.
- The air hose was connected to the designated pressure source. The valve on the pressure source was opened to initiate pressurization into the air hose. The regulator was adjusted by turning the T-screw clockwise so that a pressure was applied to the cell in 30 seconds or less. The test period begins at the time of initial pressurization.
- At the end of 30 minutes the volume of filtrate collected was measured. The air flow through the pressure regulator was shut off by turning the T-screw in a counter-clockwise direction. The valve on the pressure source was then closed and the relief valve was carefully opened.
- The assembly was then dismantled, and the mud was removed from the cup.
- The filter cake was measured using a vernier caliper, and the measurements were recorded.
- The above procedures were carried out for the other mud samples.

Hydrogen Ion Concentration (PH) - Colorimetric Paper Method

- A short strip of pH paper was placed on the surface of the sample.
- After the color of the test paper stabilized, the color of the upper side of the paper, which had not contacted the mud, was matched against the standard color chart on the side of the dispenser.
- Steps 26 and 27 were carried out on other samples.

TOXICITY TEST

After the oil based mud samples have been formulated, each is then tested on a growing plant (that is on beans seedling), to see the effects on the plant growth and the living organisms in the soil. Bean seed was planted and exposed to 100ml of three different mud samples, with the following base fluids; diesel, canola and jatropha, the growth rate was measured, and the number of days of survival.

Results of Density Measurements

The results as obtained from measurements of density using the mud balance are contained in Table 2below.

Table 2: Mud density values

SAMPLE	MEASURED DENSITY (ppg)	CALCULATED DENSITY (ppg)	ERROR	Barite (g)
Diesel	8.26	8.261	0.01	119.1
Algae	7.81	7.815	0.005	126.5
Jatropha	8.32	8.326	0.06	154.5
Moringa	8.30	8.307	0.007	149.3
Canola	8.47	8.470	0	150.6

Mud density ρ is calculated using eqn

$$\rho_m = \frac{M_{Ben} + M_{Oil} + M_{Water}}{V_{Ben} + V_{Oil} + V_{Water}}$$

e.g for Jatropha

$$\rho_{m,J} = \frac{0.110231 + 0.38040768 + 0.76742464}{0.0924608 + 0.0528344 + 0.005079769585} = 8.326 \text{ ppg}$$

(1)

From the above table, the error differences between the calculated and measured densities all lie below 0.1, thus the readings obtained using the mud balance have a high accuracy. It also showed that the denser the base oil, the higher the amount of barite needed to build.

Viscosity and Gel Strength Results

Viscosity readings obtained from the experiment carried out on the rotary viscometer are contained inTable 3.

The dial reading values (in lb/100ft²) are tabulated against the viscometer speeds in RPM.

Viscosity values are calculated with equations

Apparent viscosity= Dial Reading at 600RPM (θ_{600})/2

Table 3: Viscometer Readings for Diesel, Jatropha and Canola OBM's

Dial speed (RPM)	Diesel	Algae	Jatropha	Moringa	Canola
600	185	122	154	169	128
300	170	114	133	158	120
200	169	96	124	149	115
100	163	88	114	143	114
60	152	82	107	140	113
30	143	74	98	136	111
6	122	62	92	120	110
3	81	55	76	79	60

Table 4: Plastic Viscosities, Apparent Viscosities, Gel Strength

Rheological Properties	Diesel	Algae	Jatropha	Moringa	Canola
Plastic Viscosity	15	8	21	11	8
Apparent Viscosity	92.5	61	77	84.5	64
Gel Strength	50/51	52/43	54/55	52/53	60/72

Diesel OBM had the highest apparent viscosity, followed by Moringa, then Jatropha, Canola and algae OBM's

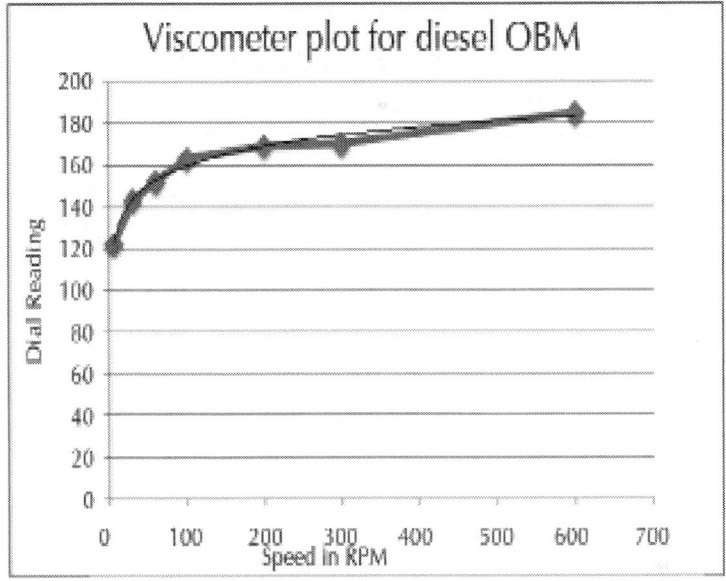

Figure 6: Viscometer Plot for Diesel OBM.

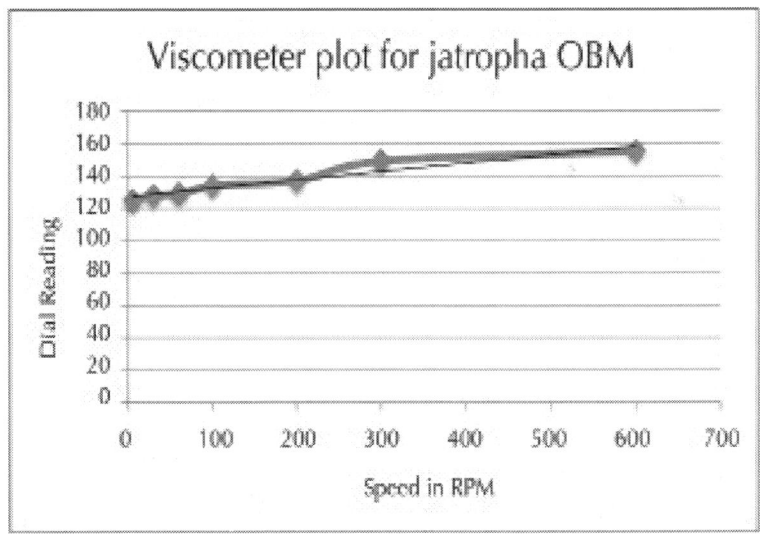

Figure 7: Viscometer Plot for Jatropha OBM.

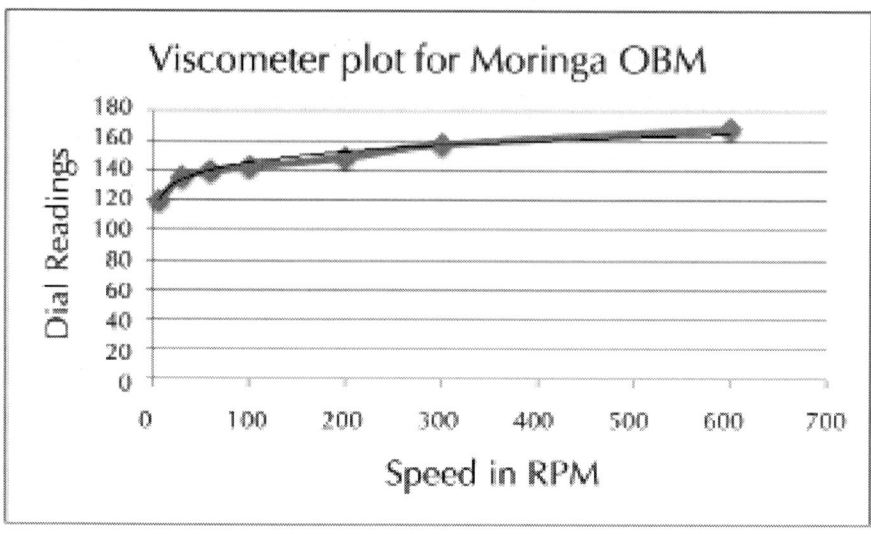

Figure 8: Viscometer Plot for Moringa OBM.

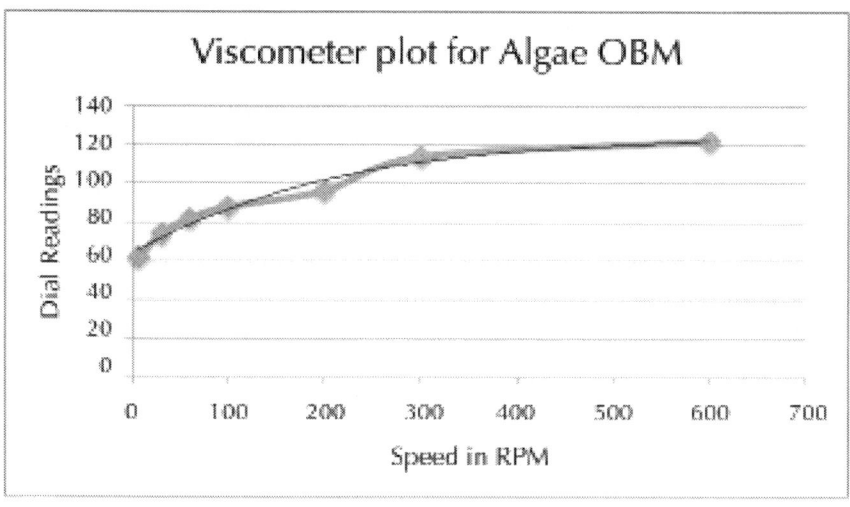

Figure 9: Viscometer Plot for algae OBM.

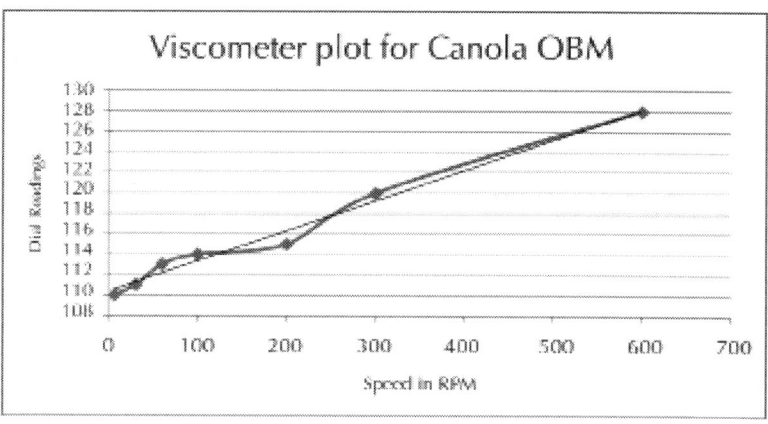

Figure 10: Viscometer Plot for Canola OBM.

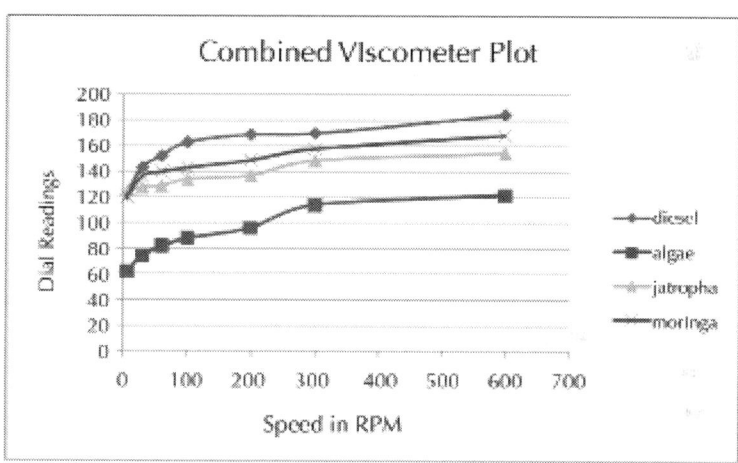

Figure 11: Combined viscometer plot for Diesel, Algae, and jatropha OBM's.

It can be seen that the plots on Figures 6 to 11, generated from the dial readings of all the mud samples are similar to the Bingham plastic model. This goes to prove that the muds have similar rheological behaviour.

However, not all the lines of the plot are as straight as the Bingham plastic model. This can be explained by a number of factors such as: possible presence of contaminants, and the possibility of behaving like a different model such as Herschel Bulkley.

A Bingham plastic fluid will not flow until the shear stress τ exceeds a certain minimum value τ_y known as the yield point[9] (Bourgoyne et al 1991). After the yield has been exceeded, the changes in shear stress are proportional to changes in shear rate and the constant of proportionality is known as the plastic viscosity μ_p.

From Figures, the yield points of the different muds can be read off. The respective yield points are the intercepts on the vertical (shear stress) axes.

For reduced friction during drilling, algae OBM gives the best results, followed by Jatropha OBM then moringa OBM.

This means Diesel OBM offers the greatest resistance to fluid flow. Algae, Jatropha, Moringa and Canola OBM's pose better prospects in the sense that their lower viscosities will mean less resistance to fluid flow. This will in turn lead to reduced wear in the drill string [10].

Mud Filtration Results

The filtration tests were carried out at 350 kPa due to the low level of the gas in the cylinder.

The mud cakes obtained from the API filter press exhibited a slick, soft texture.

From Table 5 and Figures 12 to 15, we can infer that Diesel OBM had the highest rate of filtration and spurt loss. Comparing this to a drilling scenario, this means that the mud cake from Diesel OBM is the most porous, and the thickest.

From these inferences, we can see that Algae, Jatropha, Moringa and Canola OBM's are better in filtration properties than Diesel OBM as inferred from thickness and filtration volumes.

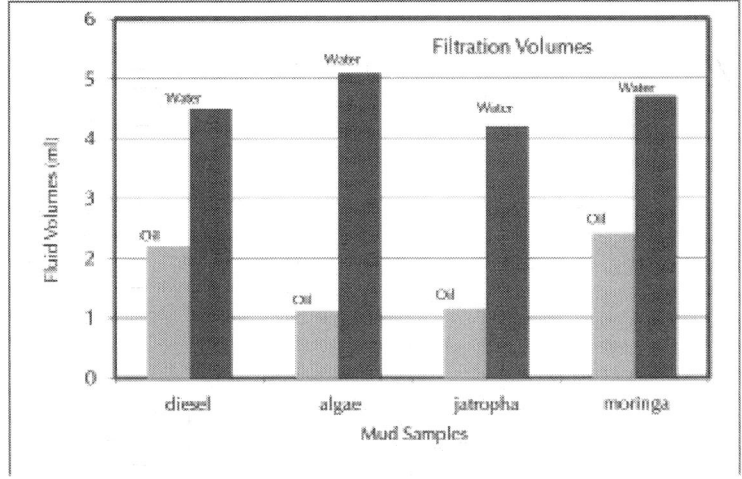

Figure 12: Filtration Volumes for Diesel, Algae, Jatropha and Moringa OBM's.

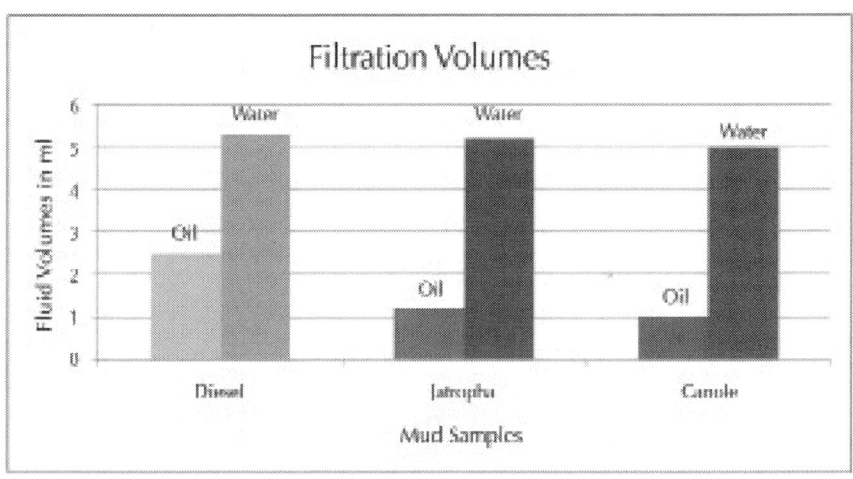

Figure 13: Filtration Volumes for Diesel, Jatropha and Canola OBM's.

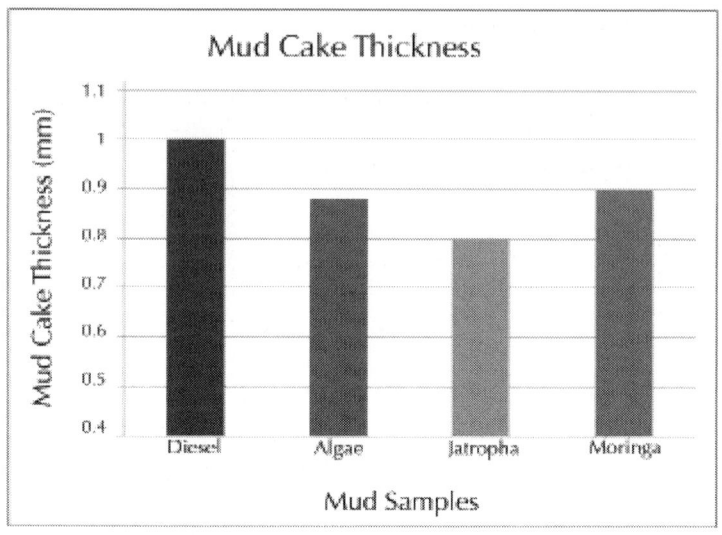

Figure 14: Mud Cake Thicknesses for Diesel, Algae, Canola OBM's.

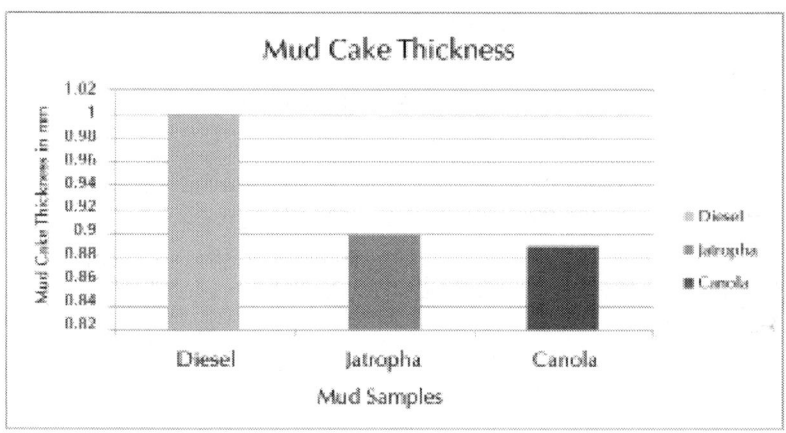

Figure 15: Mud Cake Thicknesses for Diesel, Jatropha and Canola OBM's.

Table 5: Mud Filtration Results

Filtration Properties	DIESEL	ALGAE	JATROPHA	MORINGA	Canola
Total Fluid Volume	6.9ml	6.2ml	6.3ml	7.2ml	6.0 ml
Oil volume	2.3ml	1.1ml	1.1ml	2.5ml	1.0 ml
Water Volume	4.6ml	5.1ml	4.2ml	4.7ml	4.3 ml
Cake Thickness	1.0mm	0.9mm	0.8mm	0.9mm	0.78mm

Problems caused as a result of excessive thickness include[4]:

- Tight spots in the hole that cause excessive drag.
- Increased surges and swabbing due to reduced annular clearance.
- Differential sticking of the drillstring due to increased contact area and rapid development of sticking forces caused by higher filtration rate.
- Primary cementing difficulties due to inadequate displacement of filter cake.
- Increased difficulty in running casing.

The problems as a result of excessive filtration volumes include[4]:

- Formation damage due to filtrate and solids invasion. Damaged zone too deep to be remedied by perforation or acidization. Damage may be precipitation of insoluble compounds, changes in wettability, and changes in relative permeability to oil or gas, formation plugging with fines or solids, and swelling of in-situ clays.
- Invalid formation-fluid sampling test. Formation-fluid flow tests may give results for the filtrate rather than for the reservoir fluids.
- Formation-evaluation difficulties caused by excessive filtrate invasion, poor transmission of electrical properties through thick cakes, and potential mechanical problems running and retrieving logging tools.

- Erroneous properties measured by logging tools (measuring filtrate altered properties rather than reservoir fluid properties).
- Oil and gas zones may be overlooked because the filtrate is flushing hydrocarbons away from the wellbore, making detection more difficult.

Hydrogen ION Potential Results

Drilling muds are always treated to be alkaline (i.e., a pH > 7). The pH will affect viscosity, bentonite is least affected if the pH is in the range of 7 to 9.5. Above this, the viscosity will increase and may give viscosities that are out of proportion for good drilling properties. For minimizing shale problems, a pH of 8.5 to 9.5 appears to give the best hole stability and control over mud properties. A high pH (10+) appears to cause shale problems.

The corrosion of metal is increased if it comes into contact with an acidic fluid. From this point of view, the higher pH would be desirable to protect pipe and casing (Baker Hughes, 1995).

The pH values of all the samples meet a few of the requirements stated but Diesel OBM with a pH of less than 8.5 does not meet with specification. Algae, Jatropha, Moringa and Canola OBM's show better results since their pH values fall within this range.

Table 6: pH Values

Type of Oil	DIESEL	ALGAE	JATROPHA	MORINGA
pH Value	8	9	8.5	9

Results of Cuttings Carrying Index

Only three drilling-fluid parameters are controllable to enhance moving drilled solids from the wellbore: Apparent Viscosity (AV) density (mud weight [MW]), and viscosity. Cuttings Carrying Index (CCI) is a measure of a drilling fluid's ability to conduct drilled cuttings in the hole. Higher CCI's, mean better hole cleaning capacities.

From the Table, we can see that Jatropha OBM showed best results for CCI iterations.

Table 7: Cuttings Carrying Indices (CCI's)

	Diesel	Jatropha	Canola
CCI	15.901	19.067	17.846

Pressure Loss Modeling Results

The Bingham plastic model is the standard viscosity model used throughout the industry, and it can be made to fit high shear- rate viscosity data reasonably well, and is generally associated with the viscosity of the base fluid and the number, size, and shape of solids in the slurry, while yield stress is associated with the tendency of components to build a shear-resistant.

Table 8: Bingham Plastic Pressure Losses in Psi

	Diesel	Jatropha	Canola
Drill Pipe	829	277.39	250.65
Drill Collar	177.35	173.75	157.0
Drill Collar (Open)	161.35	158.15	142.9
Drill Pipe (Open)	14.1	13.81	12.48
Drill Pipe (Cased)	9.28	9.10	8.22
Total	1191.98	706.45	571.25

It can be seen from the table that Jatropha and Canola OBM's gave better pressure loss results than Diesel OBM as a result of lower plastic viscosities, and hence should be encouraged for use during drilling activities.

Result of the Toxicity Measurements

Samples of 100ml of each of the selected oils were exposed to both corn seeds and bean seed and the no of days which the crop survived are as indicated in Figure 16. The growth rate was also measured i.e the new length of the plant was measured at regular time intervals. For the graph of toxicity of diesel based mud the reduced growth rate indicates when the leaves began to yellow, and the zero static values indicate when the plant died.

From the results indicated by the figure 16, it can be concluded that jatropha oil has less harmful effect on plant growth compared to canola and diesel. Bean seeds were planted and after one week, they were both exposed to 100ml of both jathropha formulated mud and diesel formulated mud. The seeds exposed to jatropha survived for 18 days, while that exposed to diesel mud survived for 6 days and then withered. When the soil was checked, there was no sign of any living organisms in diesel mud sample while that of the jatropha mud, there were signs of some living organisms such as earth worms, and other little insects. This shows that jatropha mud sample is environmentally safer for both plants and micro animals than diesel mud sample.

From the figure 17, it can be seen that the seeds exposed to jatropha had the highest number of days of survival which indicates its lower toxicity while that of diesel had the lowest days of survival which indicates its high toxicity. The toxicity of diesel can be traced to high aromatic hydrocarbon content. Therefore, replacements for diesel should either eliminate or minimize the aromatic contents thereby making the material non toxic or less toxic. Biodegradation and bioaccumulation however depend on the chemistry of the molecular character of the base fluids used. In general, green material i.e plant materials containing oxygen within their structure degrade easier.

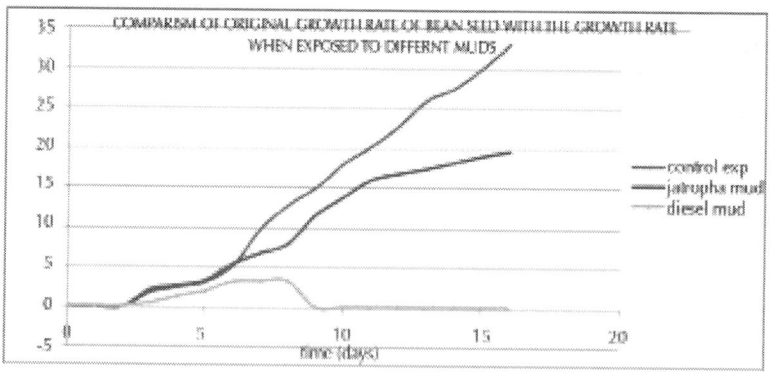

Figure 16: Comparison of Growth Rate Curve of Different Mud Types.

Results of Density Variation with Temperature

Densities were measured for the various samples at temperatures ranging from 30°C to 80°C and are summarized in Table 9.

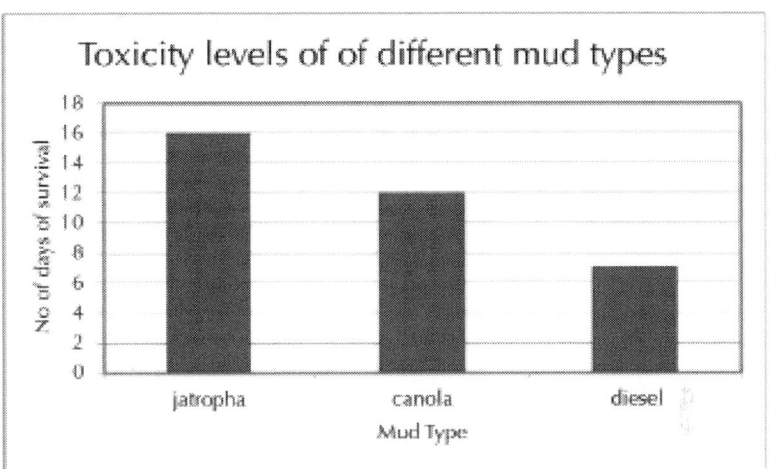

Figure 17: Toxicity of different mud types.

Table 9: Density Changes in ppg at Varying Temperatures

Temperature	Diesel	Jatropha	Canola
30°C	10	10	10
40°C	10.1	10.05	10.05
50°C	10.17	10.1	10.05
60°C	10.2	10.15	10.1
70°C	10.2	10.15	10.15
80°C	10.25	10.2	10.17

The mud samples were heated at constant pressure, and in an open system, hence the density increment.

At temperatures of 60°C and 70°C, the densities of Diesel and Jatropha OBM's were constant, while that happened with Canola OBM at a lower range of 40°C and 50°C. This is shown in Figure 18.

This could be due to the differences in temperature and heat energy required to dissipate bonds, which vary with fluid properties (i.e the continuous phases).

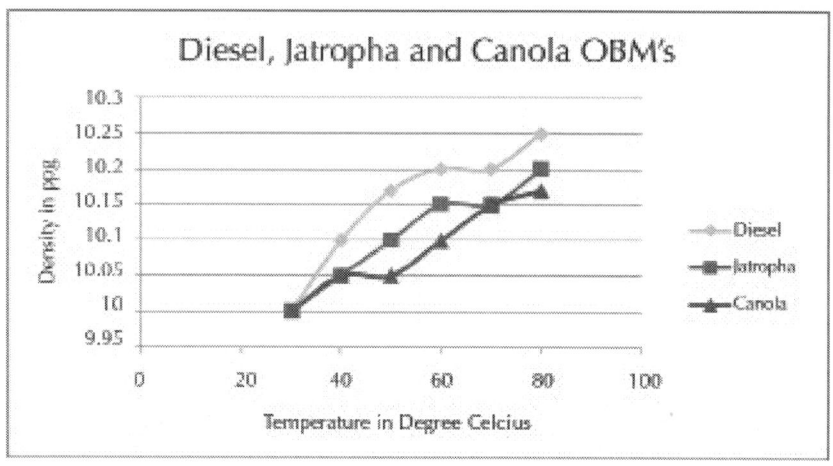

Figure 18: Density against Temperature (Diesel, Jatropha and Canola OBM's).

After the results were recorded, extrapolations were made and hypothetical values were derived for temperatures as high as 320°C, to enhance the prediction using Artificial Neural Network (ANN).

These values are summarized Tables 10 to 12

Table 10: Hypothetical Temperature-Density Values (extrapolated from regression analysis)

	Diesel	**Jatroph**a	**Canol**a
30°C	10	10	10
40°C	10.1	10.05	10.05
50°C	10.17	10.1	10.05
60°C	10.2	10.15	10.1
70°C	10.2	10.15	10.15
80°C	10.25	10.2	10.17
90°C	10.31133	10.24333	10.20667
100°C	10.35648	10.2819	10.24095

110°C	10.40162	10.32048	10.27524
120°C	10.44676	10.35905	10.30952
130°C	10.4919	10.39762	10.34381
140°C	10.53705	10.43619	10.3781
150°C	10.58219	10.47476	10.41238
160°C	10.62733	10.51333	10.44667
170°C	10.67248	10.5519	10.48095
180°C	10.71762	10.59048	10.51524
190°C	10.76276	10.62905	10.54952
200°C	10.8079	10.66762	10.58381
210°C	10.85305	10.70619	10.6181
220°C	10.89819	10.74476	10.65238
230°C	10.94333	10.78333	10.68667
240°C	10.98848	10.8219	10.72095
250°C	11.03362	10.86048	10.75524
260°C	11.07876	10.89905	10.78952
270°C	11.1239	10.93762	10.82381
280°C	11.16905	10.97619	10.8581
290°C	11.21419	11.01476	10.89238
300°C	11.25933	11.05333	10.92667
310°C	11.30448	11.0919	10.96095
320°C	11.34962	11.13048	10.99524

Results of Neural Networking

From the Artificial Neural Network Toolbox in the MATLAB 2008a, the following results were obtained:

60% of the data were used for training the network, 20% for testing, and another 20% for validation.

On training the regression values, returned values are summarized in Table 11

Table 11: Regression Values

	Diesel	**Jatropha**	**Canola**
Training	0.99999	0.99999	0.99995
Testing	0.99725	0.99056	0.99898
Validation	0.99706	0.98201	0.99328
All	0.99852	0.99414	0.99675

Since all regression values are close to unity, this means that the network prediction is a successful one.

The graphs of training, testing and validation are presented below:

The values were returned after performing five iterations for each network. This also goes to say that the Artificial Neural Network, after being trained and simulated, is a viable and feasible instrument for prediction.

Figures 19 to 31 present the plots of Experimental data against Estimated (predicted) data for training, testing and validation processes from MATLAB 2008.

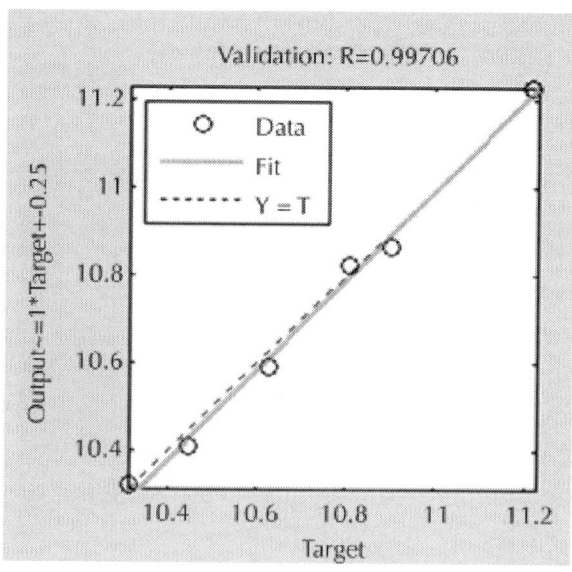

Figure 19: Diesel OBM Validation values.

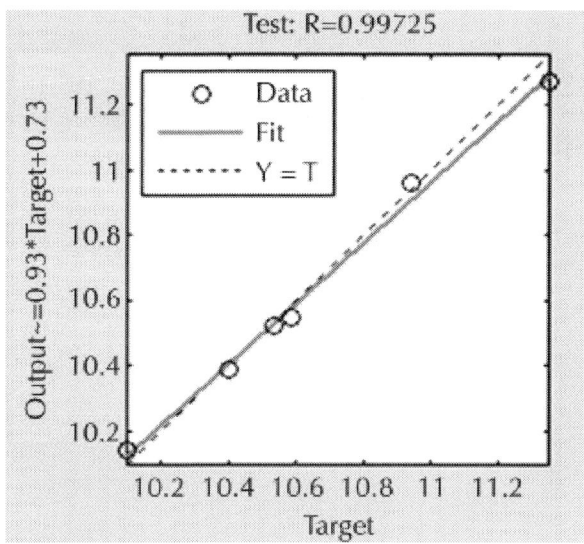

Figure 20: Diesel OBM Test values.

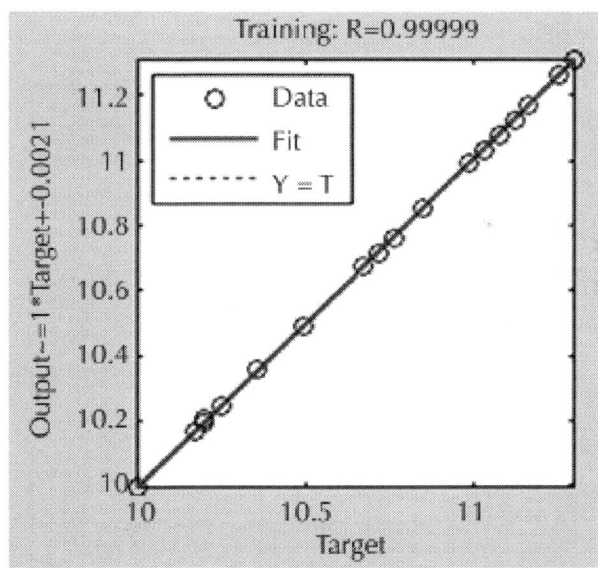

Figure 21: Diesel OBM Training values.

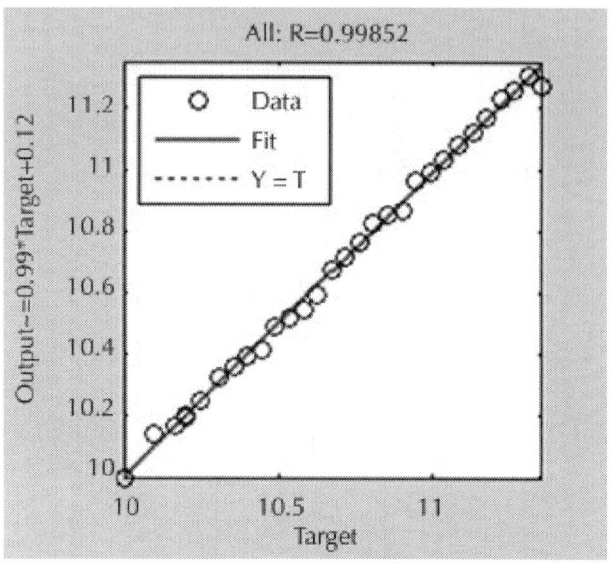

Figure 22: Diesel OBM Overall values.

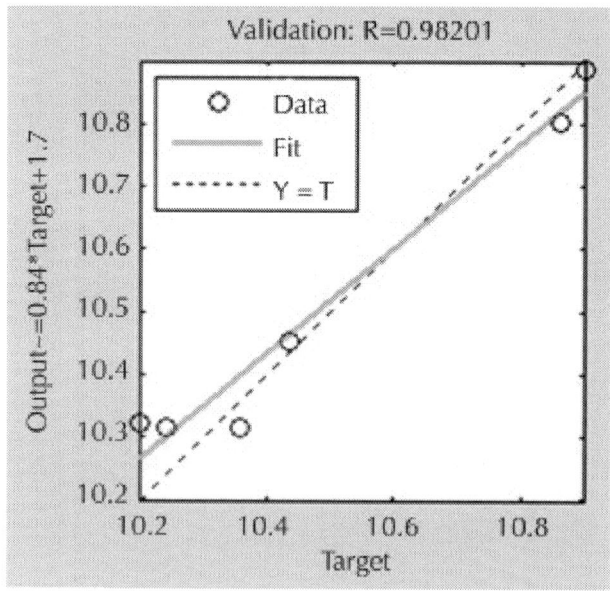

Figure 23: Diesel OBM Overall values.

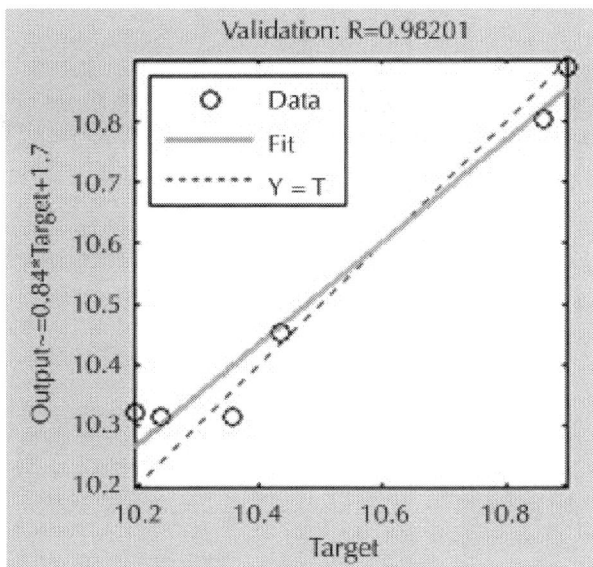

Figure 24: Jatropha OBM Validation values.

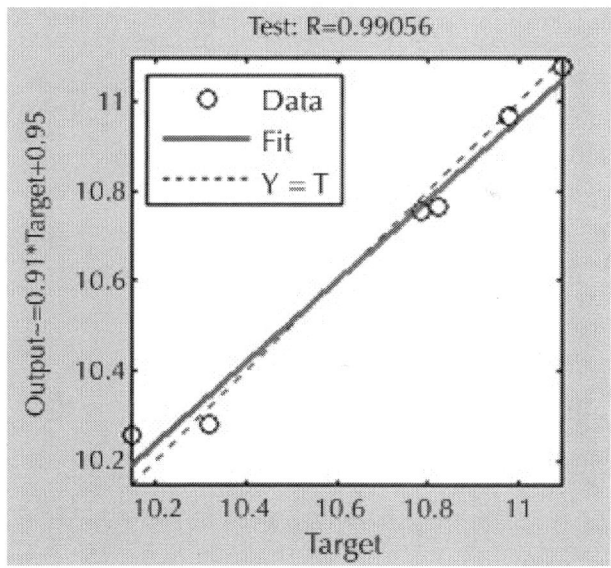

Figure 25: Jatropha OBM Test values.

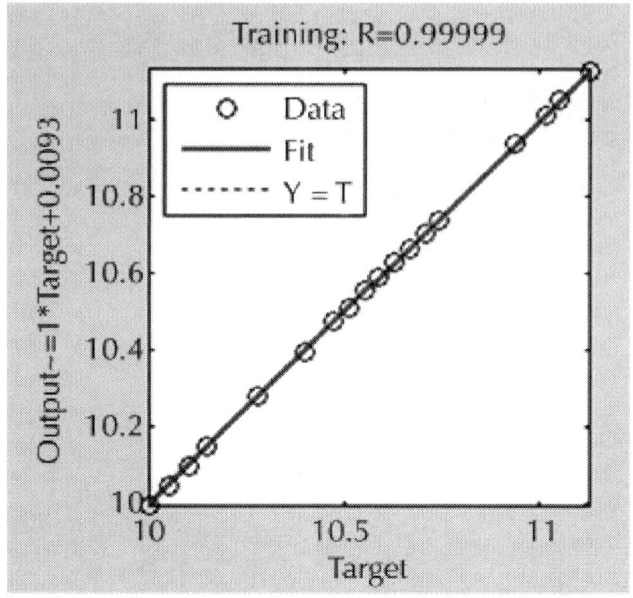

Figure 26: Jatropha OBM Training values.

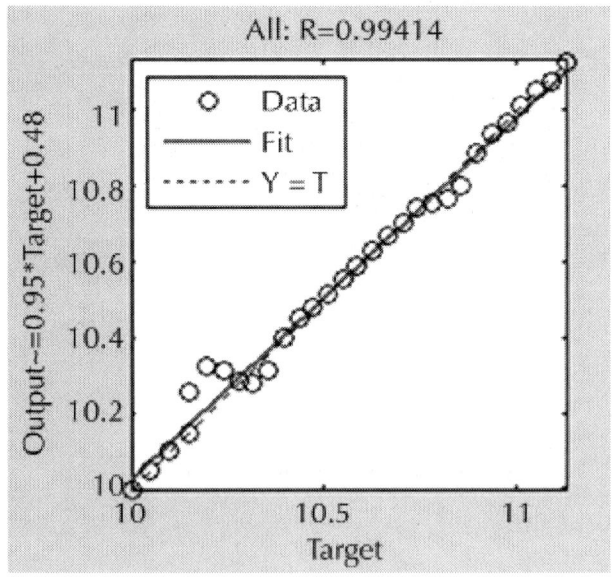

Figure 27: Jatropha OBM Overall values.

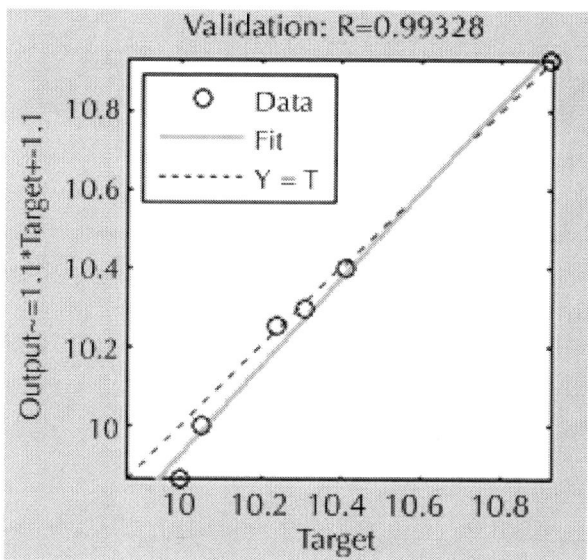

Figure 28: Canola OBM Validation values.

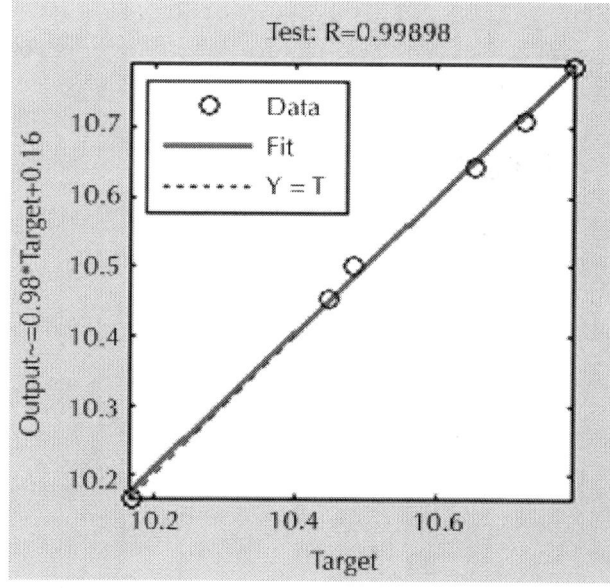

Figure 29: Canola OBM Test values.

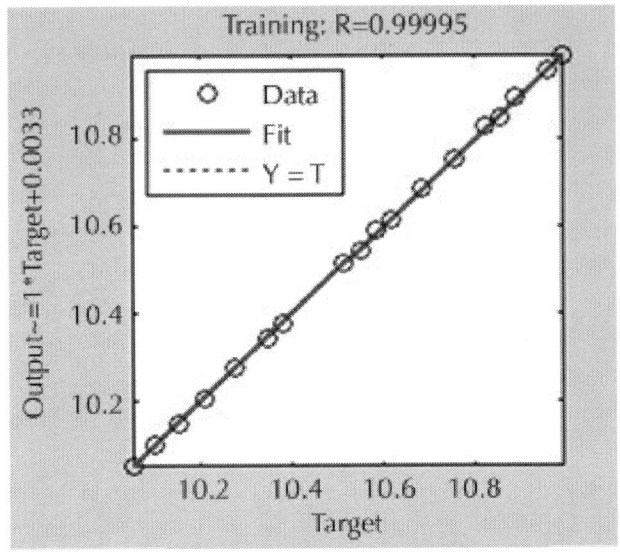

Figure 30: Canola OBM Training values.

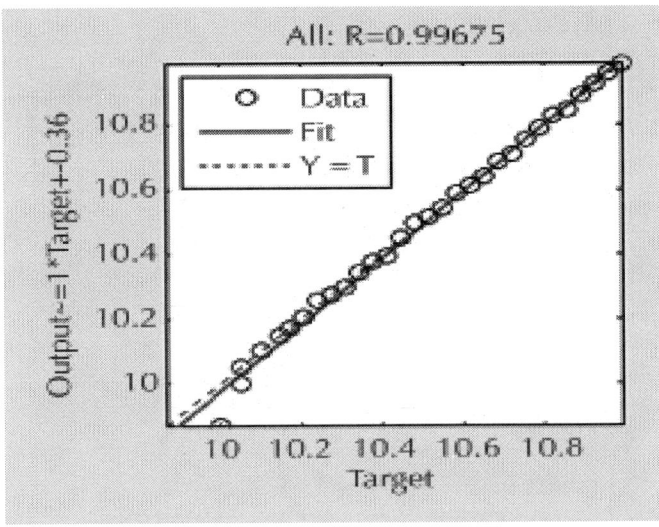

Figure 31: Canola OBM Overall values.

We can see from the Figures 19 to 31 that the data points all align closely with the imaginary/arbitrary straight line drawn across. This validates the accuracy of the network predictions and this also gives rise to the high regression values (tending towards unity) presented in Table 11

Errors, estimated values and experimental values are summarized in Tables 12 to 14

Table 12: Errors, Experimental Values, and Estimated Values for Diesel OBM

Temperature °C	Exp Values	Est Values	Errors
30	10	10.049	0.049
40	10.1	10.1407	0.0407
50	10.17	10.1794	0.0094
60	10.2	10.2022	0.0022
70	10.2	10.2236	0.0236
80	10.25	10.24	-0.01
90	10.31133	10.287	-0.02433
100	10.35648	10.3579	0.001424
110	10.40162	10.3904	-0.01122
120	10.44676	10.4222	-0.02456
130	10.4919	10.4835	-0.0084
140	10.53705	10.5204	-0.01665
150	10.58219	10.5455	-0.03669
160	10.62733	10.6133	-0.01403
170	10.67248	10.687	0.014524
180	10.71762	10.7202	0.002581
190	10.76276	10.7714	0.008638
200	10.8079	10.8335	0.025595
210	10.85305	10.8611	0.008052
220	10.89819	10.8991	0.00091
230	10.94333	10.9623	0.018967
240	10.98848	10.9955	0.007024
250	11.03362	11.0273	-0.00632
260	11.07876	11.085	0.006238
270	11.1239	11.1195	-0.0044

280	11.16905	11.1474	-0.02165
290	11.21419	11.2049	-0.00929
300	11.25933	11.2432	-0.01613
310	11.30448	11.2545	-0.04998
320	11.34962	11.2674	-0.08222

Table 13: Errors, Experimental Values, and Estimated Values for Jatropha OBM

Temperature ºC	Exp Values	Est Values	Errors
30	10	10	0
40	10.05	10.05	0
50	10.1	10.0998	-0.0002
60	10.15	10.1485	-0.0015
70	10.15	10.2556	0.1056
80	10.2	10.3232	0.1232
90	10.24333	10.3143	0.070967
100	10.2819	10.2851	0.003195
110	10.32048	10.281	-0.03948
120	10.35905	10.3147	-0.04435
130	10.39762	10.3985	0.000881
140	10.43619	10.4526	0.01641
150	10.47476	10.4769	0.002138
160	10.51333	10.5126	-0.00073
170	10.5519	10.5544	0.002495
180	10.59048	10.5884	-0.00208
190	10.62905	10.63	0.000952
200	10.66762	10.6665	-0.00112
210	10.70619	10.7025	-0.00369
220	10.74476	10.741	-0.00376
230	10.78333	10.7559	-0.02743
240	10.8219	10.7655	-0.0564
250	10.86048	10.803	-0.05748
260	10.89905	10.8872	-0.01185

270	10.93762	10.9375	-0.00012
280	10.97619	10.9644	-0.01179
290	11.01476	11.0148	3.81E-05
300	11.05333	11.0533	-3.3E-05
310	11.0919	11.0747	-0.0172
320	11.13048	11.1305	2.38E-05

Table 14: Errors, Experimental Values, and Estimated Values for Canola OBM

Temperature °C	Exp Values	Est Values	Errors
30	10	9.8841	-0.1159
40	10.05	10.0044	-0.0456
50	10.05	10.048	-0.002
60	10.1	10.0925	-0.0075
70	10.15	10.1449	-0.0051
80	10.17	10.1681	-0.0019
90	10.20667	10.1987	-0.00797
100	10.24095	10.2489	0.007948
110	10.27524	10.2745	-0.00074
120	10.30952	10.2972	-0.01232
130	10.34381	10.3445	0.00069
140	10.3781	10.377	-0.0011
150	10.41238	10.4003	-0.01208
160	10.44667	10.4539	0.007233
170	10.48095	10.4994	0.018448
180	10.51524	10.519	0.003762
190	10.54952	10.5537	0.004176
200	10.58381	10.5952	0.01139
210	10.6181	10.6145	-0.0036
220	10.65238	10.6444	-0.00798
230	10.68667	10.6888	0.002133
240	10.72095	10.7105	-0.01045
250	10.75524	10.7365	-0.01874

260	10.78952	10.7895	-2.4E-05
270	10.82381	10.8224	-0.00141
280	10.8581	10.8465	-0.0116
290	10.89238	10.8971	0.004719
300	10.92667	10.9337	0.007033
310	10.96095	10.945	-0.01595
320	10.99524	10.9562	-0.03904

The minute errors encountered in the predictions further justify the claim that the ANN is a trust worthy prediction tool.

The Experimental outputs were then plotted against their corresponding temperature values, and also fitted into the polynomial trend line of order 2.

The Equations derived are[7]:

Diesel OBM:

$$\rho = -4 \times 10^{-7}T^2 + 0.004T + 9.915$$

(2)

Jatropha OBM:

$$\rho = 7 \times 10^{-7}T^2 + 0.003T + 9.994$$

(3)

Canola OBM:

$$\rho = -2 \times 10^{-6}T^2 + 0.004T + 9.827$$

(4)

Also by comparing the networks created with that of Osman and Aggour[12] (2003), we can see that this work is technically viable in predicting mud densities at varying temperatures as the network developed in the course of this project showed regression values close to those proposed by Osman and Aggour [12].

Errors, percentage errors and average errors as compared with Osman and Aggour[12] are relatively lower, thus guaranteeing the accuracy of the newly modeled network.

Table 15 shows the regression values of Osman and Aggour for oil based mud density variations with temperature and pressure [12].

Table 15: Table Showing the Regression Values from Osman and Aggour [12]

Training	Testing	Validation	All
0.99978	0.99962	0.99979	0.9998

Table 16: Table of the Relative Deviations

Temperature	Diesel	Jatropha	Canola
30	0.49	0	1.159
40	0.40297	0	0.453731
50	0.092429	0.00198	0.0199
60	0.021569	0.014778	0.074257
70	0.231373	1.040394	0.050246
80	0.097561	1.207843	0.018682
90	0.235986	0.692808	0.078054
100	0.013748	0.031076	0.077606
110	0.107859	0.382504	0.007183
120	0.235115	0.428105	0.119538
130	0.080107	0.008473	0.006675
140	0.157991	0.157237	0.010553
150	0.346719	0.020412	0.116025
160	0.132049	0.006975	0.069241
170	0.136087	0.023647	0.176011
180	0.024081	0.019604	0.035776
190	0.080259	0.00896	0.039587
200	0.23682	0.01049	0.107622
210	0.074195	0.03447	0.03386
220	0.008346	0.035012	0.074922
230	0.173317	0.254405	0.019963
240	0.06392	0.521209	0.097495

250	0.057271	0.529223	0.174223
260	0.056307	0.108703	0.000221
270	0.039597	0.001088	0.013022
280	0.193818	0.107419	0.106789
290	0.082846	0.000346	0.043324
300	0.143289	0.000302	0.064369
310	0.442092	0.155111	0.145538
320	0.724421	0.000214	0.355045

Table 17 compares the Average Absolute Percent Error abbreviation (AAPE), Maximum Average relative deviation (Ei) and Minimum E_i for Diesel, Jatropha and Canola OBM's as well as the values from Osman and Aggour.

Table 17: Table Comparing Maximum Ei, Minimum Ei, and AAPE

	Diesel	Jatropha	Canola	Osman et al
Minimum $_E i$	0.008346	0.000214	0.000221	0.102269
Maximum $_E i$	0.724421	1.207834	1.159	1.221067
AAPE	0.172738	0.193426	0.124949	0.36037

CONCLUSIONS

The lower viscosities of jatropha, moringa and canola oil based mud (OBM's) make them very attractive prospects in drilling activities.

The results of the tests carried out indicate that jatropha, moringa and canola OBM's have great chances of being among the technically viable replacements of diesel OBM's. The results also show that additive chemistry must be employed in the mud formulation, to make them more technically feasible. In addition, the following conclusions were drawn:

- From the viscosity test results, it can be inferred that the plastic viscosity of jatropha OBM can be further stepped down by adding an adequate concentration of thinner. This method can also be used to reduce the gel strengths of jatropha, moringa and canola OBM's.

- The formulated drilling fluids exhibited Bingham plastic behavior, and from the pressure loss modeling, canola OBM gave the best results, and next was jatropha OBM.
- The tests of temperature effects on density: The densities increased and became constant at some point, and began increasing again (these temperature points of constant density varied for the different samples). The diesel OBM showed the highest variation range, while the canola OBM showed the lowest.
- Artificial Neural Network works well for prediction of scientific parameters, due to minimized errors returned.

LIMITATIONS

- The temperature-density tests were carried out at surface conditions under an open system and at a constant pressure due to the absence of a pressure unit thus, the equations developed are not guaranteed for down-hole circulating conditions.
- During the temperature-density tests, it was observed that some of the mud particles settled at the base of the containing vessel, and this reduced the accuracy of the readings.
- The accuracy of the temperature-density readings is also reduced because of the use of an analogue mud balance (calibrated to the nearest 0.1 ppg).
- The mud samples were aged for only 24 hours, hence the feasibility of older muds may not be guaranteed.

RECOMMENDATIONS

- This work should further be tested and investigated for the effect of temperature on other properties of the formulated drilling fluids.
- The temperature-density tests should also be carried out at varying pressures, to simulate downhole conditions.

ACKNOWLEDGEMENTS

We wish to thank all members of staff Department of Petroleum Engineering Covenant University, Nigeria for their technical support in carrying out this research work especially Mr Daramola. We also acknowledge the support of Environmental Research Group, Father-Heroes Forte Technology Nigeria for their commitment.

REFERENCES

1.   A. Yassin, A. Kamis, O. A. Mohamad, 1991Formulation of an Environmentally Safe Oil Based Drilling Fluid" SPE 23001, Paper presented at the SPE Asia Pacific Conference held in Perth, Western Australia, 47November 1991.

2.   Terry Hemphil,1996Prediction of Rheological Behavior of Ester-Based Drilling Fluids Under Downhole Conditions" SPE 35330. Paper presented at.s1 Ihe 1996 SPE International Petroleum Conference and Exhibition of Mexico held in Villahermosa, Tabasco 57March 1996.

3.   A.M. Ezzat and K.A. Al-Buraik,1997Environmentally Acceptable Drilling Fluids for Offshore Saudi Arabia" SPE 37718. Paper presented at the SPE Middle East Oil Show and Conference held in Bahrain, 1518March, 1997.

4.   G. Sáchez, N. León, M. Esclapés, I. Galindo, A. Martínez, J. Bruzual, I. Siegert, 1999Environmentally Safe Oil-Based Fluids for Drilling Activities" SPE 52739, Paper presented at SPE/EPA Exploration and Production Environmental Conference held in Austin, Texas, 28 February-3 March 1999.

5.   H. Xiaoqing, Z. Lihui, 2009Research on the Application of Environment Acceptable Natural Macromolecule Based Drilling Fluids" SPE 123232, Paper presented at the SPE Asia Pacific Health, Safety, Security and Environment Conference and Exhibition held in Jarkata, Indonesia, 46August 2009.

6.   A. Dosunmu, J. Ogunrinde, 2010Development of Environmentally Friendly Oil Based Mud using Palm Oil and Groundnut Oil". SPE 140720. Paper presented at the 34th Annual International

Conference and Exhibition in Tinapa-Calabar, Nigeria, July 31st-August 7th, 2010.

7. *Fadairo Adesina, Adeyemi Abiodun, Ameloko Anthony, Falode Olugbenga, (SLO2012Modelling the Effect of Temperature on Environmentally Safe Oil Based Drilling Mud Using Artificial Neural Network Algorithm" Journal of Petroleum and Coal 2012, 541

8. *Fadairo Adesina, Ameloko Anthony, Adeyemi Gbadegesin, Ogidigbo Esseoghene, Airende Oyakhire2012Environmental Impact Evaluation of a Safe Drilling Mud" SPE Middle East Health, Safety, Security, and Environment Conference and Exhibition held in Abu Dhabi, UAE, 24April 2012, SPE-152865-PP

9. T. A. Bourgoyne, K. K. Millheim, M. E. Chenevert, ". Applied, Engineering". S. P. E. Drilling, series. Textbook, Vol, 1991

10. Mitchell B., Advanced Oil Well Drilling Engineering Hand Book, Mitchell Engineering, 10th edition, 1995.

11. Baker Hughes Mud Engineering Hand Book

12. Osman, E.A. and Aggour, M.A.: "Determination of Drilling Mud Density Change with Pressure and Temperature Made Simple and Accurate by ANN," paper SPE 81422 presented at the2003SPE Middle East Oil Show and Conference, Bahrain, 58April.

Chapter 6

High-Efficiency Separation of Bio-Oil

Shurong Wang[1]

[1]Zhejiang University, China

INTRODUCTION

What is Fast Pyrolysis?

Biomass is a CO_2-neutral energy source that has considerable reserve. It can replace fossil feedstock in the production of heat, electricity, transportation fuels, chemicals, and various materials. Liquid bio-fuels, which are considered to be substitutes for traditional petrol liquid fuels, can be produced from biomass in different ways, such as high-pressure liquefaction, hydrothermal pyrolysis, and fast pyrolysis.

Fast pyrolysis is a technology that can efficiently convert biomass feedstock into liquid biofuels. The liquid obtained from fast pyrolysis, which is also called crude bio-oil, may be used as burning oil in boilers or even as a transportation fuel after upgrading. Fast pyrolysis is a process in which lignocellulosic molecules of biomass are rapidly decomposed to short-chain molecules in the absence of oxygen. Under conditions of high heating rate, short residence time, and moderate pyrolysis temperature, pyrolysis vapor and some char are generated. After condensation of the pyrolysis vapor, liquid product can be collected in a yield of up to 70 wt% on a dry weight basis (Bridgwater et al., 1999; Lu et al., 2009). The obvious advantages of the process are as follows:

- Low-grade biomass feedstock can be transformed into liquid biofuels with relatively higher heating value, thus making storage and transportation more convenient.

- The by-products are char and gas, which can be used to provide the heat required in the process or be collected for sale.

- For waste treatment, fast pyrolysis offers a method that can avoid hazards such as heavy metal elements in the char and reduce pollution of the environment.

Many researchers have focused on the techniques of fast pyrolysis, and various configurations of reactor have been developed to satisfy the requirements of high heating rate, moderate reaction temperature, and short vapor residence time for maximizing bio-oil production. During the past decades, many types of reactor have been designed to promote the large-scale and commercial utilization of biomass fast pyrolysis, such as the fluidized bed reactor (Luo et al., 2004; Wang et al., 2002), the ablative reactor (Peacocke & Bridgwater, 1994), the rotating cone reactor (Muggen, 2010;Peacocke; Wagenaar, 1994) and Vacuum reactor (Bridgwater, 1999; Yang et al., 2001).

The Composition and Properties of Bio-Oil

The chemical composition of bio-oil is significantly different from that of petroleum fuels. It consists of different compounds derived from decomposition reactions of cellulose, hemicellulose, and lignin. The chemical composition of bio-oil varies depending on the type of biomass feedstock and the operating parameters. Generally speaking, bio-oil is a mixture of water and complex oxygen-rich organic compounds, including almost all such kinds of organic compounds, that is, alcohols, organic acids, ethers, esters, aldehydes, ketones, phenols, etc. Normally, the component distribution of bio-oil may be measured by GC-MS analysis.

Crude bio-oil derived from lignocellulose is a dark-brown, viscous, yet free-flowing liquid with a pungent odor. Crude bio-oil has an oxygen content of 30−50 wt%, resulting in instability and a low heating value (Oasmaa & C., 2001). The water content of bio-oil ranges from 15 to 50 wt%. The high water content of bio-oil derives from water in the feedstock and dehydration reactions during biomass pyrolysis (Bridgwater, 2012). Heating value is an important indicator for fuel oils. The heating value of bio-oil is usually lower than 20 MJ/kg, much lower than that of fuel oil. The high water content and oxygen content are two factors responsible for its low heating value. The density of bio-oil derived from fast pyrolysis is within the range 1100−1300 kg/m^3 (Adjaye et al., 1992). The pH value of bio-oil is usually in the range 2−3 owing to the presence of carboxylic acids such as formic acid and acetic acid. The strong acidity can corrode pipework and burner components. Measurements of the corrosiveness of bio-oil have shown that it can induce an apparent mass loss of carbon steel and the breakdown of a diesel engine burner (Wright et al., 2010).

Fresh bio-oil is a homogeneous liquid containing a certain amount of solid particles. After long-term storage, it may separate into two layers and heavy components may be deposited at the bottom. As mentioned above, the high content of oxygen and volatile organic compounds are conducive to the ageing problems of bio-oil. The aldol condensation of aldehydes and alcohols and self-aggregation of aldehydes to oligomers are two of the most likely reactions to take place. Coke and inorganic components in the bio-oil may also have a catalytic effect, thereby enhancing the ageing process (Rick & Vix, 1991).

The Utilization of Bio-Oil

The oxygenated compounds in bio-oil can lead to several problems in its direct combustion, such as instability, low heating value, and high corrosiveness. Although higher water content can improve the flow properties and reduce NOx emissions in the fuel combustion process, it causes many more problems. It not only decreases the heating value of the fuel, but also increases the corrosion of the combustor and can result in flame-out. The low pH value of bio-oil also aggravates corrosiveness problems, which may lead to higher storage and transportation costs. Many researchers have tested the combustion of bio-oil in gas boiler systems, diesel engines, and gas turbines (Czernik & Bridgwater, 2004).

Fresh bio-oil from different feedstocks can generally achieve stable combustion in a boiler system. One problem, however, is the difficulty of ignition. The high water content of bio-oil not only decreases its heating value, but also consumes a large amount of latent heat of vaporization (Bridgwater & Cottam, 1992). Thus, the direct ignition of bio-oil in a cold furnace is not easy, and an external energy source is needed for ignition and pre-heating of the furnace. The combustion of bio-oil in diesel engines is more challenging. Its long ignition delay time, short burn duration, and lower peak heat release have limited its combustion properties (Vitolo & Ghetti, 1994). Experiments employing bio-oil in gas turbines have proved largely unsuccessful. The high viscosity and high ash content of bio-oil result in severe blocking and attrition problems in the injection system. Moreover, acid in the bio-oil is harmful to the mechanical components of the gas turbine.

Even though many combustion tests of bio-oil have shown its combustion performances to be inferior to those of fossil fuels, the environmental advantages of bio-oil utilization cannot be ignored. Comparative tests have shown that the SO_2 emissions from bio-oil combustion are much lower than those from fossil fuel combustion.

Bio-oil is a mixture of many organic chemicals, such as acetic acid, turpentine, methanol, etc. Many compounds in bio-oil are important chemicals, such as phenols used in the resins industry, volatile organic acids used to produce de-icers, levoglucosan, hydroxyacetaldehyde, and some agents applied in the pharmaceutical, synthetic fiber, and fertilizer industries, as well as flavoring agents for food products (Radlein, 1999). Besides, bio-oil can also be used in a process that converts traditional lime into bio-lime (Dynamotive Corporation, 1995).

SEPARATION OF BIO-OIL FOR UPGRADING OR REFINEMENT

The Importance of Separation Technology

Bio-oil cannot be directly applied as a high-grade fuel because of its inferior properties, such as high water and oxygen contents, acidity, and low heating value. Thus, it is necessary to upgrade bio-oil to produce a high-grade liquid fuel that can be used in engines (Bridgwater, 1996; Czernik & Bridgwater, 2004; Mortensen et al., 2011).

In view of its molecular structure and functional groups, and using existing chemical processes for reference, such as hydrodesulfurization, catalytic cracking, and natural gas steam reforming, several generic bio-oil upgrading technologies have been developed, including hydrogenation, cracking, esterification, emulsification, and steam reforming.

Components with unsaturated bonds, such as aldehydes, ketones, and alkenyl compounds, influence the storage stability of bio-oil, and hydrogenation could be used to improve its overall saturation (Yao et al., 2008). Hydrogenation can achieve a degree of deoxygenation of about 80%, and transform bio-oil into high-quality liquid fuel (Venderbosch et al., 2010; Wildschut et al., 2009). This process requires a high pressure of hydrogen, which increases both the complexity and cost of the operation. Alcohol hydroxyl, carbonyl, and carboxyl groups were easily hydrodeoxygenated, and phenol hydroxyl and ether groups were also reactive, while furans, having a cyclic structure, were more difficult to convert (Furimsky, 2000). After the separation of bio-oil, the components with alcohol hydroxyl, carbonyl, carboxyl, phenol hydroxyl, and ether groups can be efficiently hydrodeoxygenated at a low hydrogen pressure, while the hydrodeoxygenation of more complex components, such as ethers and furans, may be achieved by developing special catalysts.

Catalytic cracking of bio-oil refers to the reaction whereby oxygen is removed in the form of CO, CO_2, and H_2O, in the presence of a solid acid catalyst, such as zeolite, yielding a hydrocarbon-rich high-grade liquid fuel. In the process of cracking, oxygenated compounds in bio-oil are thought to undergo initial deoxygenation to form light olefins, which are then cyclized to form aromatics or undergo some other reactions to produce hydrocarbons (Adjaye & Bakhshi, 1995a). Since bio-oil has a relatively low H/C ratio, and dehydration is accompanied by the loss of hydrogen, the H/C ratio of the final product is generally low, and carbon deposits with large aromatic structures tend to be formed, which can lead to deactivation of the catalyst (Guo et al., 2009a). The cracking of crude bio-oil is always terminated in a short time, with a coke yield of about 20% (Adjaye & Bakhshi, 1995b; Vitolo et al., 1999). Alcohols, ketones, and carboxylic acids are efficiently converted into aromatic hydrocarbons, while aldehydes tend to condense to form carbon deposits (Gayubo et al., 2004b). Phenols also show low reactivity and coking occurs readily (Gayubo et al., 2004a). Besides, some thermally sensitive compounds, such as pyrolitic lignin, might undergo aggregation to form a precipitate, which would block the reactor and lead to deactivation of the catalyst. Consequently, efforts have been made to avoid this phenomenon by separating these compounds through thermal pre-treatment (Valle et al., 2010). Therefore, to maintain the stability and high performance of the cracking process, it is necessary to obtain fractions suitable for cracking by separation of bio-oil, to achieve the partial conversion of bio-oil into hydrocarbon fuels.

Bio-oil has a high content of carboxylic acids, so catalytic esterification is used to neutralize these acids. Both solid acid and base catalysts display high activity for the conversion of carboxylic acids into the corresponding esters, and the heating value of the upgraded oil is thereby increased markedly (Zhang et al., 2006). Since this method is more suitable for the transformation of carboxylic acids, which constitute a relatively small proportion of crude bio-oil, an ester fuel with a high heating value can be expected to be produced from the esterification of a fraction enriched with carboxylic acids obtained from the separation.

The emulsion fuel obtained from bio-oil and diesel is homogeneous and stable, and can be burned in existing engines. Research on the production of emulsions from crude bio-oil and diesel suggested that the emulsion produced was more stable than crude bio-oil. Subsequent tests of these emulsions in different diesel engines showed that because of the presence of carboxylic acids, the injector nozzle was corroded, and this corrosion was accelerated by the high-velocity turbulent flow in the spray channels (Chiaramonti et al., 2003a; Chiaramonti et al., 2003b). Besides corrosion, the high water content of bio-oil will lower the heating value of the emulsion as a fuel, and some high molecular weight components such as sugar oligomers and pyrolitic lignin will increase the density and reduce the volatility of the emulsion. Thus, it is beneficial to study the emulsification of the separated fractions that contain less water and fewer high molecular weight components.

Catalytic steam reforming of bio-oil is also an important upgrading technology for converting it into hydrogen. Research on the steam reforming of acetic acid and ethanol is now comparatively mature, with high conversion of reactants, hydrogen yields, and stability of the catalysts (Hu & Lu, 2007). However, some oxygenated compounds in bio-oil show inferior reforming behavior. Phenol cannot be completely converted even at a high steam-to-carbon ratio, while m-cresol and glucose not only show low reactivity, but are also easily coked (Constantinou et al., 2009; Hu & Lu, 2009). To improve the reforming process, some further investigations of steam reforming based on other separating methods are needed.

Therefore, it is necessary to combine crude bio-oil utilization with the current upgrading technologies. Taking advantage of efficient bio-oil separation to achieve the enrichment of compounds in the same family or the components that are suitable for the same upgrading method is a significant strategy for the future utilization of high-grade bio-oil.

Conventional Separation Technologies

The efficient separation of bio-oil establishes a solid foundation for its upgrading. Currently, conventional methods for bio-oil separation include column chromatography, solvent extraction, and distillation.

Solvent Extraction

The solvents for extraction include water, ethyl acetate, paraffins, ethers, ketones, and alkaline solutions. In recent years, some special solvents, such as supercritical CO_2, have also been used for extraction or other research. By selecting appropriate solvents for extraction of the desired products, good separation of bio-oil can be achieved.

Some researchers have used non-polar solvents for the primary separation of bio-oil, such as toluene and n-hexane, and then proceeded to extract the solvent-insoluble fraction with water; finally, the water-soluble and water-insoluble fractions were further extracted with diethyl ether and dichloromethane, respectively (Garcia-Perez et al., 2007; Oasmaa et al., 2003). A lot of organic solvents are consumed during the process. Considering the cost of these solvents and the difficulty of the recovery process, the operating costs are unacceptable, which hinders its industrialization.

Supercritical fluid extraction is based on the different dissolving abilities of supercritical solvents under different conditions. Supercritical fluid extraction at low temperatures contributes to preventing undesirable reactions of thermally sensitive components. Researchers usually use CO_2 as the supercritical solvent. In a supercritical CO_2 extraction, compounds of low polarity (aldehydes, ketones, phenols, etc.) are selectively extracted, while acids and water remain in the residue phase (Cui et al., 2010).

Column Chromatography

The principle of column chromatography is that substances are separated based on their different adsorption capabilities on a stationary phase. In general, highly polar molecules are easily adsorbed on a stationary phase, while weakly polar molecules are not. Thus, the process of column chromatography involves adsorption, desorption, re-adsorption, and re-desorption. Silica gel is commonly used as the stationary phase, and an eluent is selected according to the polarity of the components. Paraffin eluents, such as hexane and pentane, are used to separate aliphatic compounds. Aromatic compounds are usually eluted with benzene or toluene. Some other polar compounds are obtained by elution with methanol or other polar solvents (Ertas & Alma, 2010; Onay et al., 2006;Putun et al., 1999).

Distillation

Distillation is a common separating technology in the chemical industry. This method separates the components successively according to their different volatilities, and it is essential for the separation of liquid mixtures. Atmospheric pressure distillation, vacuum distillation, steam distillation, and some other types of distillation have been applied in bio-oil separation.

Due to its complex composition, the boiling of bio-oil starts below 100 °C under atmospheric pressure, and then the distillation continues up to 250−280 °C, whereupon 35−50% of residue is left (Czernik & Bridgwater, 2004).

The thermal sensitivity of bio-oil limits the operating temperature of distillation. In view of the unsatisfactory results obtained by atmospheric pressure distillation, researchers have employed vacuum distillation to lower the boiling points of components, and bio-oil could thereby be separated at a low temperature. Characterization of the distilled organic fraction showed that it had a much better quality than the crude bio-oil, containing little water and fewer oxygenated compounds, and having a higher heating value.

Steam distillation is performed by introducing steam into the distilling vessel, to heat the bio-oil and decrease its viscosity, and finally the volatile components are expelled by the steam. In a study combining steam distillation with reduced pressure distillation, bio-oil was first steam distilled to recover 14.9% of a volatile fraction. The recovered fraction was then further distilled by reduced pressure distillation to recover 16 sub-fractions (Murwanashyaka et al., 2001). In this process, a syringol-containing fraction was separated and syringol with a purity of 92.3% was obtained.

Due to its thermal sensitivity, it is difficult to efficiently separate bio-oil by conventional distillation methods. Molecular distillation seems to offer a potential means of realizing bio-oil separation, because it has the advantages of low operating temperature, short heating time, and high separation efficiency.

Molecular Distillation

There are forces between molecules, which can be either repulsive or attractive depending on intermolecular spacing. When molecules are close together, the repulsive force is dominant. When molecules are not very close to each other, the forces acting between them are attractive in nature, and there should be no intermolecular forces if the distance between molecules is very large. Since the distances between gas molecules are large, the intermolecular forces are negligible, except when molecules collide with each other. The distance between collisions with another molecule is called its free path.

The mean free path of an ideal gas molecule can be described by Eq. (1):

$$\lambda_m = \frac{k}{\sqrt{2}\pi} \frac{T}{d^2 p}$$

(1)

Where T (°C) is the local temperature; $_m$ (m) refers to the mean free path; d (m) is the effective diameter of the molecule; P (Pa) is the local pressure; and k is the Boltzmann constant.

As is apparent from Eq. (1), the molecular mean free path is inversely proportional to the pressure and the square of the effective molecular diameter. Under certain conditions, that is, if the temperature and pressure are fixed, the mean free path is a function of the effective molecular diameter. Apparently, a smaller molecule has a shorter mean free path than a larger molecule. Furthermore, molecular mean free path will increase with increasing temperature or decreasing pressure.

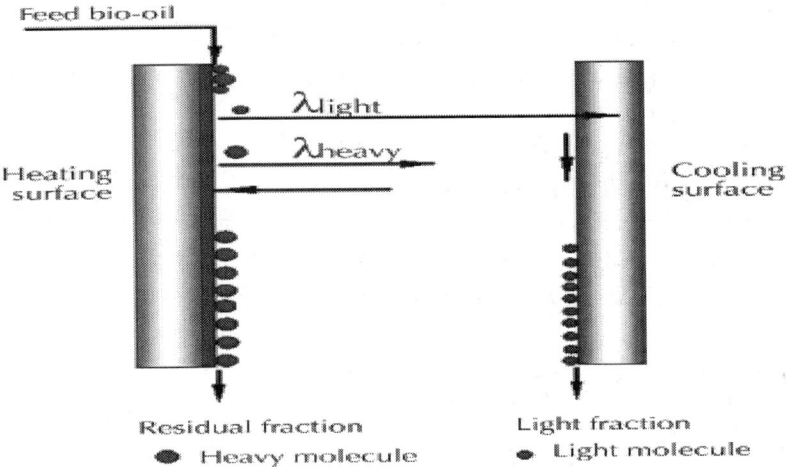

Figure 1: Schematic representation of molecular distillation.

Molecules will move more rapidly when the liquid mixture is heated. Surface molecules will overcome intermolecular forces and escape as gas molecules when they obtain sufficient energy. With an increased amount of gas molecules above the liquid surface, some molecules will return to the surface. Under certain conditions, the molecular motion will achieve dynamic equilibrium, which is manifested as equilibrium on a macroscopic scale.

Traditional distillation technology separates components by differences in their boiling points. However, molecular distillation (or short-path distillation) is quite different and precisely relies on the various mean free paths of different substances. As shown in Fig. 1, the distance between the cooling and heating surfaces is less than the mean free path for a light molecule, but greater than that for a heavy molecule. Therefore, the light molecules escaping from the heating surface can easily reach the cooling surface and be condensed. The dynamic balance is thereby broken, and the light molecules are continuously released from the liquid phase. On the contrary, the heavy molecules are not released and return to the liquid phase. In this way, the light and heavy molecules are effectively separated.

Molecular distillation technology has been widely used in the chemical, pharmaceutical, and foodstuff industries, as well as in scientific research to concentrate and purify organic chemicals. It is a feasible process for the separation of thermally unstable materials, taking into account that it only takes a few seconds to complete the separation process. Bio-oil is a complex mixture of many compounds with a wide range of boiling points. It is thermally sensitive and easily undergoes reactions such as decomposition, polymerization, and oxygenation. Additionally, most of the compounds are present in low concentrations. Molecular distillation is not limited by these unfavorable properties and is suitable for the separation of bio-oil to facilitate analysis and quantification of its constituent compounds.

HIGH-EFFICIENCY SEPARATION OF BIO-OIL AT ZHEJIANG UNIVERSITY

A Molecular Distillation Apparatus

Fig. 2 shows a KDL-5 wiped-film molecular distillation apparatus used for bio-oil separation research at Zhejiang University, which was manufactured by UIC Corporation in Germany. It consists of four main units, namely a feeding unit, an evaporation unit, a condensation unit, and a reduced pressure unit. The feeding unit mainly comprises a graduated dosing funnel with a double jacket, which is filled with heat-transfer oil to control the temperature and to ensure free flowing of the feedstock. The evaporation unit comprises a cylindrical evaporator with a surface area of 0.048 m², encased in a double jacket containing heat-transfer oil to maintain good temperature homogeneity. It is worth noting that all of the temperatures of these sections are independent. The condensation unit has two cold traps. The first cold trap (or internal condenser) is located in the center of evaporator, and condenses the volatile compounds reaching the cooling surface. There is another cold trap to prevent uncondensed volatile organic compounds from entering the pump. In the reduced pressure unit, the condensation temperature is usually set at −25 °C. The evaporation temperature ranges from room temperature to 250 °C, while the operating pressure can be as low as 5 Pa.

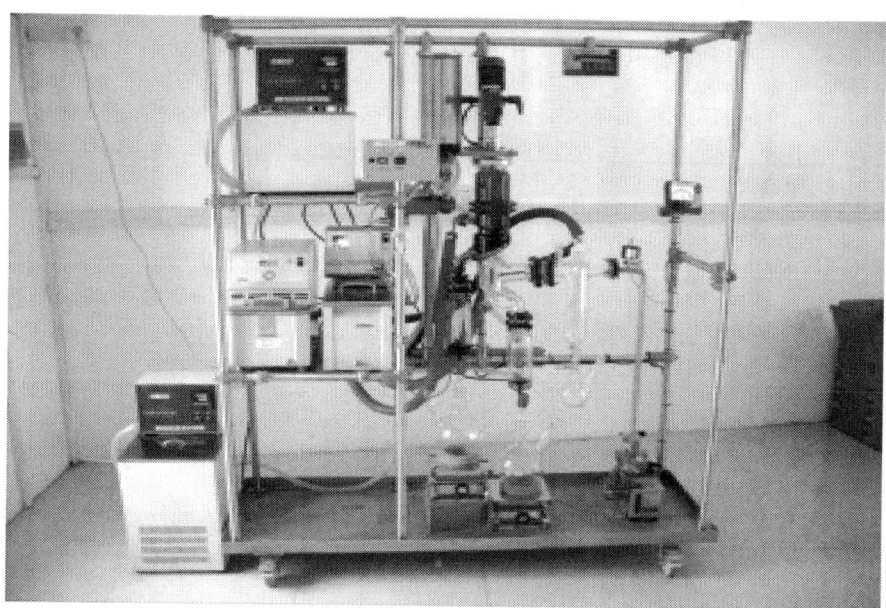

Figure 2: KDL-5 molecular distillation apparatus.

The bio-oil used at Zhejiang University was produced from a bench-scale fluidized bed fast pyrolysis reactor (Wang et al., 2008). Crude bio-oil often contains some solid particles, which would abrade the evaporator surface and block the orifice of the dosing funnel, so it is necessary to perform some pre-treatments. Centrifugation and filtration are usually used to remove the solid particles, and traditional reduced pressure distillation can also be used to remove water and volatile compounds. The pre-treated bio-oil is placed in the funnel and then the separation process starts. The volatile components released from the thin liquid film are condensed by the internal condenser to form the distilled fraction, while the heavy compounds that are not vaporized flow along the evaporator surface and are collected as the residual fraction.

Because of the short residence time of the feed material at the evaporation temperature, this gentle distillation process only puts a low thermal load on the materials to be distilled. It is therefore appropriate for the separation of bio-oil, which is thermally unstable.

Single Separation Process under Different Operating Conditions

Physical Characteristics of Samples

Bio-oil used in the single separation process was produced by the pyrolysis of Mongolian pine sawdust (Wang et al., 2008). Wang et al. (Guo et al., 2009b; Wang et al., 2009) carried out experimental research on molecular separation of the bio-oil, which was pre-treated by centrifugation and filtration to remove solid particles. Molecular distillation of the bio-oil at 50, 70, 100, and 130 °C, respectively, was investigated under a fixed pressure of 60 Pa. Under all of the tested conditions, the light fraction collected by the second condenser placed before vacuum pump was designated as LF, the middle fraction condensed by the internal condenser as MF, and the heavy fraction as HF.

The color of the distilled fractions becomes lighter while the residual fractions become darker. Under the four conditions, water was concentrated in the LFs, which had water contents of about 70 wt%. The LFs could not be burned because of their high water contents. The pH values of the LFs were in the range 2.13−2.17 as a result of their carboxylic acid contents. On the other hand, the HFs had the highest heating values and the lowest water contents, resulting in good ignitability but inferior fluidity. At a distillation temperature of 70 °C, the water content of the MF was as low as 2 wt%. The total mass of the bio-oil distillation fractions amounted to more than 97% of the bio-oil feed. With increasing temperature, the yield of the LF increased without any coking or polymerization problem. Water and volatile carboxylic acids were evaporated from the feedstock in the temperature range 50−130 °C under low pressure, and more carboxylic acids escaped from the liquid at higher temperature. However, on further increasing the temperature, this phenomenon was not so pronounced, due to more and more molecules of higher boiling point also being distilled. The yield of the distilled fraction increased with increasing distillation temperature. However, too high temperature may lead to decomposition of some chemical compounds in the crude bio-oil. Hence, there must be an optimum temperature to realize reasonable separation.

Distribution of Acidic Compounds in Bio-Oil Fractions

The high content of carboxylic acids in bio-oil is one of the main reasons for its corrosiveness, which damages storage tanks, boilers, and gas turbines. As a consequence, detailed research on the separation of acidic compounds has been carried out under the condition of distillation at 50 °C.

The carboxylic acid content in the refined bio-oil was used to estimate the separation efficiency. Guo et al. (2009b) chose five major acids in bio-oil and studied their separation characteristics. As shown inFig. 3, the amount of acetic acid, the most abundant acid in bio-oil, was reduced to 1.9 wt% and 0.96 wt% in the MF and HF, respectively. The results indicated that acidic compounds could be effectively separated from the crude bio-oil by means of molecular distillation technology. The LF, which was rich in water and carboxylic acids, was valuable for further catalytic esterification of bio-oil acidic compounds. Both MF and HF could be further upgraded to produce high-quality fuels.

Distribution Characteristics of Several Chemicals in Three Fractions

Fig. 4 illustrates the distributions of selected compounds in bio-oil, MF, and HF. Six chemicals were selected as being representative of ketones, aldehydes, phenols, and sugars, respectively. 1-Hydroxy-2-propanone, the most abundant ketone in bio-oil, could not be detected in the MF or HF after separation, indicating that it was extremely enriched in the LF. The content of furfural in the HF was just a little higher than that in the MF, but much lower than that in bio-oil. The distributions of these two compounds reflected the enrichment of small ketone and aldehyde molecules in the LF. Phenol appeared to be difficult to separate as there were similar distributions in bio-oil, MF, and HF. In contrast, compounds of higher molecular weight tended to be enriched in the MF and HF. For example, 1,2-benzenediol and 3-methyl-1,2-benzenediol were more abundant in the MF and HF than in the bio-oil before separation. In particular, the relative content of 1,2-benzenediol amounted to 11.73 wt% in HF, about five times higher than that in bio-oil (2.45 wt%).

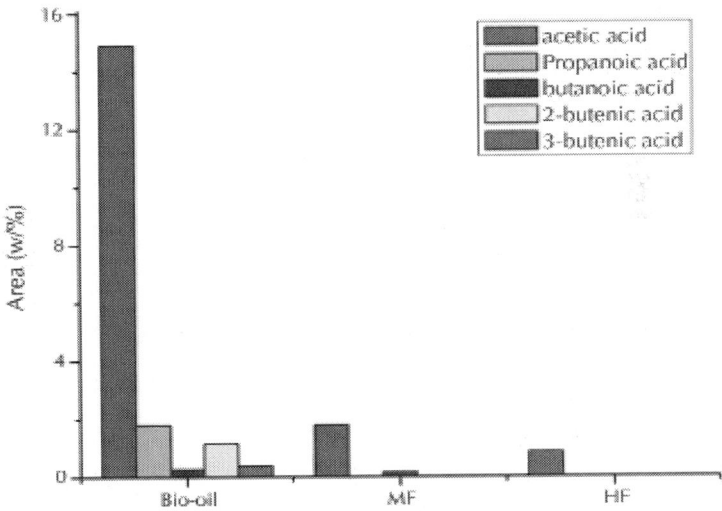

Figure 3: Contents of acidic compounds in three samples obtained at 50 °C (Guo et al., 2009b).

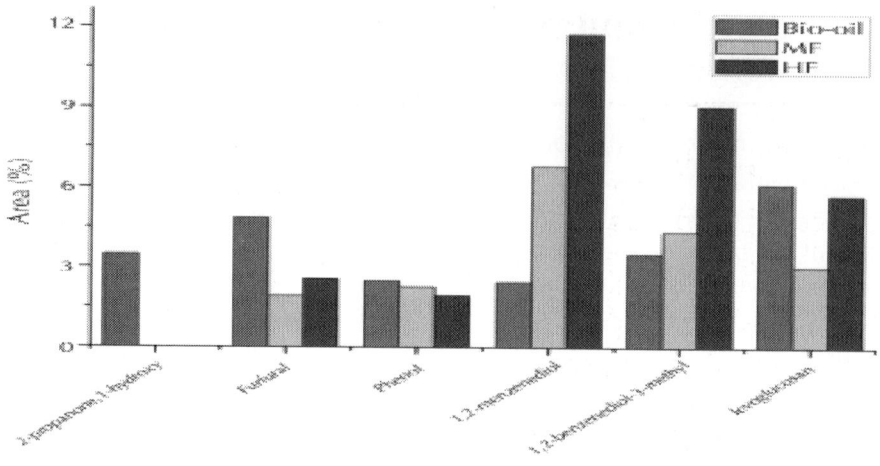

Figure 4: Distributions of selected compounds obtained at 50 °C in three fractions.

Statistical Method to Evaluate the Separation of Bio-Oil

As the composition of the bio-oil and the effects of the operating conditions on the distribution of each fraction are both complicated, Wang et al. (2009) put forward a statistical method to directly evaluate the separation level of bio-oil by molecular distillation. The separation coefficients of four groups, "Complete Isolation", "Nonvaporization", "Enrichment", and "Even Distribution", were calculated from the ratios of relative peak of a single component with respect to total components. The results showed that "Complete Isolation" had the largest percentage, followed by "Even Distribution", "Non-vaporization", and "Enrichment" which contained only small parts. Meanwhile, the temperature had a significant effect on the distributions of the compounds.

Multiple Molecular Distillation for Bio-Oil Separation

Based on the above single distillation experiments, a multiple molecular distillation experiment was carried out to further evaluate the separation characteristics of bio-oil (Guo et al., 2010b). The feed bio-oil, which was pre-treated by centrifugation, filtration, and vacuum distillation, was firstly distilled at 80 °C and 1600 Pa to obtain the distilled fraction 1 (DF-1) and the residual fraction 1 (RF-1). A part of RF-1 was then further distilled at 340 Pa to obtain DF-2 and RF-2 fractions. In the multiple distillation process, the distilled fraction yield of each distillation process was about 26 wt%. The amounts of water in RF-1 and RF-2 were greatly reduced. The RFs from the two processes had higher heating values than the feed bio-oil or DFs. The acid content was 11.37 wt% in the feed bio-oil, while it was 17.36 wt% for DF-1, nearly four times higher than that in RF-1 (4.56 wt%). In the second process, the acid content of RF-2 was further reduced to 1.38 wt%. The content of monophenols in RF-1 was 36.24 wt%, about twice that in DF-1 (18.02 wt%). Sugars showed non-distillable character in the two distillation processes, and no amounts could be detected in the DF.

In order to gain a deeper insight into the bio-oil distillation properties, Guo (Guo et al., 2010b) proposed a separation factor to evaluate the separation characteristics. The separation factors of acetic acid and 1-hydroxy-2-propanone were approximately 0.9, implying that they could be mostly distilled off. 2-Methoxyphenol, phenol, 2(5H)-furanone, and 2-methoxy-4-methylphenol, the separation factors of which ranged from 0.61 to 0.74, proved to be difficult to separate effectively. Higher molecular weight compounds, such as 3-methoxy-1,2-benzenediol, 4-methoxy-1,2-benzenediol, and 1,2-benzenediol, were very difficult to distil, having separation factors close to zero.

The Joint Distillation System at Zhejiang University

Based on the operation experiences gained with the KDL5 molecular distillation apparatus, a larger-scale joint reduced pressure and molecular distillation set-up was established in the State Key Laboratory of Clean Energy Utilization, Zhejiang University. The flow diagram of this joint distillation system is illustrated in Fig. 5. The processing capacities of the reduced pressure distillation and molecular distillation units were both 8–10 kg/h, and they could be run at temperatures up to 300 °C and pressures down to 50 Pa. The reduced pressure distillation unit could be operated separately to remove the water from bio-oil as well as to obtain bio-oil fractions. When these two units were assigned to run together, the pre-treated bio-oil from the first reduced pressure distillation unit could be pumped directly into the molecular distillation unit.

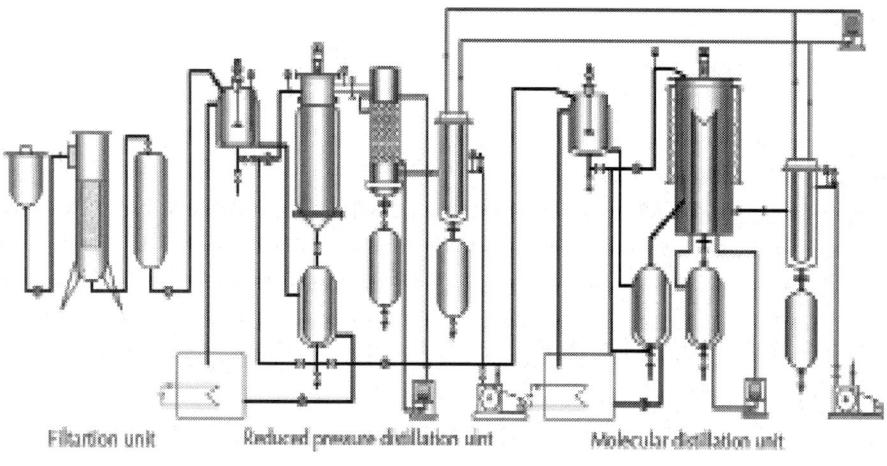

Filtration unit   Reduced pressure distillation uint   Molecular distillation unit

Figure 5: Schematic diagram of the joint distillation system.

Further Research on the Distilled Fractions

Based on the molecular distillation results, a scheme of the process combining molecular distillation separation with bio-oil upgrading is proposed. The light fraction rich in carboxylic acids and other light components could be used for esterification, catalytic cracking, and steam reforming, to produce ester fuel, hydrocarbons, and hydrogen, respectively. For the middle fraction, steam reforming at high temperature or hydrodeoxygenation at high pressure could efficiently convert this fraction into hydrogen or hydrocarbons. The heavy fraction, which consisted mainly of pyrolytic lignin and sugar oligomers, could be emulsified with diesel to obtain emulsion fuel with a relatively high heating value. On the other hand, the extraction of some valuable chemicals can benefit the overall economy of this process.

Recently, some further research has been performed, aiming at investigating some characteristics of the distilled fractions and devising more promising upgrading methods. Thermal decomposition processes and the pyrolysis products of crude bio-oil and distilled fractions were investigated by means of TG-FTIR by Guo (Guo et al., 2010a). The light fraction (LF) was completely evaporated at 30−150 °C, with the maximum weight loss rate at about 100 °C due to the volatilization of water and compounds of lower boiling point. The middle fraction (MF) and heavy fraction (HF) contained more lignin-derived compounds, and these decomposed continuously over a wide temperature range of 30−600 °C, leaving a final residue yield of 25−30%. Upgrading of the distilled fraction rich in carboxylic acids and ketones was carried out by Guo (Guo et al., 2011). Carboxylic acids accounted for 18.39% of the initial fraction, with acetic acid being the most abundant. After upgrading, the carboxylic acid content decreased to 2.70%, with a conversion yield of 85.3%. The content of esters in the upgraded fraction increased dramatically from 0.72% to 31.1%. The conversion of corrosive carboxylic acids into neutral esters reduced the corrosivity of the bio-oil fraction.

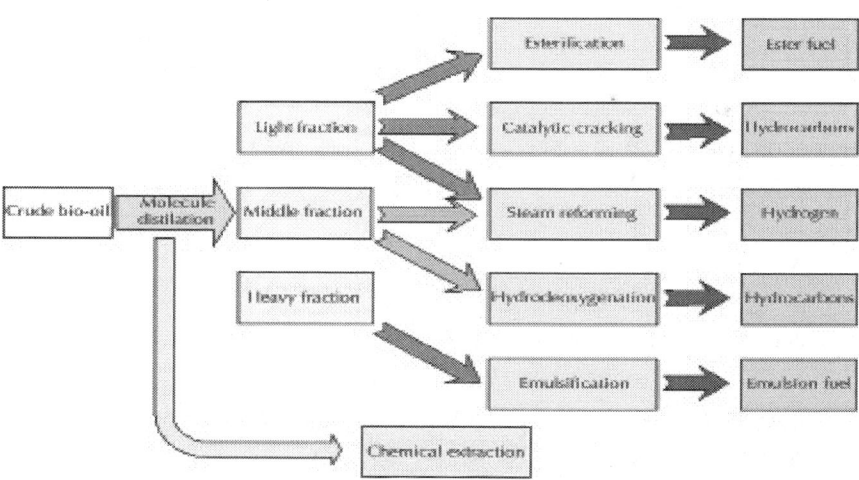

Figure 6: A scheme of the process combining molecular distillation separation with bio-oil upgrading.

ACKNOWLEDGEMENT

The author acknowledges the financial support from the Program for New Century Excellent Talents in University, the International Science & Technology Cooperation Program of China (2009DFA61050), Zhejiang Provincial Natural Science Foundation of China (R1110089), the Research Fund for the Doctoral Program of Higher Education of China (20090101110034), the National Natural Science Foundation of China (50676085) and the National High Technology Research and Development Program of China (2009AA05Z407). The author also highly appreciates the kind support from Mr. Zuogang Guo, Mr. Qinjie Cai, Mr. Long Guo and Miss Yurong Wang, who have been involved in the experimental research and the preparation of this chapter.

REFERENCES

1. J. D. Adjaye, N. N. Bakhshi, 1995aCatalytic Conversion Of A Biomass-Derived Oil To Fuels And Chemicals.1. Model-Compound Studies And Reaction Pathways. Biomass & Bioenergy, 831311490961-9534

2. J. D. Adjaye, N. N. Bakhshi, 1995bProduction Of Hydrocarbons By Catalytic Upgrading Of A Fast Pyrolysis Bio-Oil.1. Conversion Over Various Catalysts. Fuel Processing Technology, 4531611830378-3820

3. J. D. Adjaye, R. K. Sharma, N. N. Bakhshi, 1992Characterization And Stability Analysis Of Wood-Derived Bio-Oil. Fuel Processing Technology, 3132412560378-3820

4. A. V. Bridgwater, 1999Principles and practice of biomass fast pyrolysis processes for liquids. Journal of Analytical and Applied Pyrolysis, 511-23220165-2370

5. A. V. Bridgwater, 1996Production of high grade fuels and chemicals from catalytic pyrolysis of biomass. Catalysis Today, 291-42852950920-5861

6. A. V. Bridgwater, 2012Review of fast pyrolysis of biomass and product upgrading. Biomass and Bioenergy, 3868940961-9534

7.   A. V. Bridgwater, M. L. Cottam, 1992Opportunities For Biomass Pyrolysis Liquids Production And Upgrading. Energy & Fuels, 621131200887-0624

8.   A. V. Bridgwater, D. Meier, D. Radlein, 1999An overview of fast pyrolysisofbiomass.OrganicGeochemistry,3012147914931479-1493

9.   D. Chiaramonti, A. Bonini, E. Fratini, 2003aDevelopment of emulsions from biomass pyrolysis liquid and diesel and their use in engines- Part 1: emulsion production. Biomass & Bioenergy, 25185990961-9534

10. D. Chiaramonti, A. Bonini, E. Fratini, 2003bDevelopment of emulsions from biomass pyrolysis liquid and diesel and their use in engines- Part 2: tests in diesel engines. Biomass & Bioenergy, 2511011110961-9534

11. D. A. Constantinou, J. L. G. Fierro, A. M. Efstathiou, 2009The phenol steam reforming reaction towards H2 production on natural calcite. Applied Catalysis B-Environmental, 903-43473590926-3373

12. H. Cui, J. Wang, S. Wei, 2010Supercritical CO2 extraction of bio-oil. Journal Of Shandong Unirersity Of Technology (Science And Technology), 246151672-6197

13. S. Czernik, A. V. Bridgwater, 2004Overview of applications of biomass fast pyrolysis oil. Energy & Fuels, 1825905980887-0624

14. Dynamotive Corporation.1995Acid emission reduction.U.S. Patent 5458803.

15. M. Ertas, M. H. Alma, 2010Pyrolysis of laurel (Laurus nobilis L.) extraction residues in a fixed-bed reactor: Characterization of bio-oil and bio-char. Journal of Analytical and Applied Pyrolysis, 88122290165-2370

16. E. Furimsky, 2000Catalytic hydrodeoxygenation. Applied Catalysis A-General, 19921471900092-6860X.

17. M. Garcia-Perez, A. Chaala, H. Pakdel, 2007Characterization of bio-oils in chemical families. Biomass & Bioenergy, 3142222420961-9534

18. A. G. Gayubo, A. T. Aguayo, A. Atutxa, 2004aTransformation of oxygenate components of biomass pyrolysis oil on a HZSM-

5 zeolite. I. Alcohols and phenols. Industrial & Engineering Chemistry Research, 4311261026180888-5885

19. A. G. Gayubo, A. T. Aguayo, A. Atutxa, 2004bTransformation of oxygenate components of biomass pyrolysis oil on a HZSM-5 zeolite. II. Aldehydes, ketones, and acids. Industrial & Engineering Chemistry Research, 4311261926260888-5885

20. X. J. Guo, S. R. Wang, Z. G. Guo, 2010aPyrolysis characteristics of bio-oil fractions separated by molecular distillation. Applied Energy, 879289228980306-2619

21. X. Y. Guo, Y. Zheng, B. H. Zhang, 2009aAnalysis of coke precursor on catalyst and study on regeneration of catalyst in upgrading of bio-oil. Biomass & Bioenergy, 3310146914730961-9534

22. Z. G. Guo, S. R. Wang, Y. L. Gu, 2010bSeparation characteristics of biomass pyrolysis oil in molecular distillation. Separation and Purification Technology, 76152571383-5866

23. Z. G. Guo, S. R. Wang, G. H. Xu, 2011Upgrading Of Bio-Oil Molecular Distillation Fraction With Solid Acid Catalyst. Bioresources, 63253925501930-2126

24. Z. G. Guo, S. R. Wang, Y. Y. Zhu, 2009bSepatation of acid compounds for refining biomass pyrolysis oil. Journal of Fuel Chemistry and Technology, 37149521872-5813

25. X. Hu, G. X. Lu, 2007Investigation of steam reforming of acetic acid to hydrogen over Ni-Co metal catalyst. Journal of Molecular Catalysis A-Chemical, 261143481381-1169

26. X. Hu, G. X. Lu, 2009Investigation of the steam reforming of a series of model compounds derived from bio-oil for hydrogen production. Applied Catalysis B-Environmental, 883-43763850926-3373

27. Q. Lu, W. Z. Li, X. F. Zhu, 2009Overview of fuel properties of biomass fast pyrolysis oils. Energy Conversion and Management, 505137613830196-8904

28. Q. Lu, X. L. Yang, X. F. Zhu, 2008Analysis on chemical and physical properties of bio-oil pyrolyzed from rice husk. Journal of Analytical and Applied Pyrolysis, 8221911980165-2370

29. Z. Y. Luo, S. R. Wang, Y. F. Liao, 2004Research on biomass fast pyrolysis for liquid fuel. Biomass & Bioenergy, 2654554620961-9534

30. G. Muggen, 2010Empyro Project Summary. in: PyNe newsletter, 2735

31. J. N. Murwanashyaka, H. Pakdel, C. Roy, 2001Seperation of syringol from birch wood-derived vacuum pyrolysis oil. Separation and Purification Technology, 241-21551651383-5866

32. A. Oasmaa, E. Kuoppala, Y. Solantausta, 2003Fast pyrolysis of forestry residue. 2. Physicochemical composition of product liquid. Energy & Fuels, 1724334430887-0624

33. O. Onay, A. F. Gaines, O. M. Kockar, 2006Comparison of the generation of oil by the extraction and the hydropyrolysis of biomass. Fuel, 8533823920016-2361

34. G. V. C. Peacocke, Techno-economic assessment of power production from the Wellman and BTG fast pyrolysis. Science in Thermal and Chemical Biomass Conversion, 2006217851902

35. G. V. C. Peacocke, A. V. Bridgwater, 1994Ablative Plate Pyrolysis Of Biomass For Liquids. Biomass & Bioenergy, 71-61471540961-9534

36. A. E. Putun, A. Ozcan, E. Putun, 1999Pyrolysis of hazelnut shells in a fixed-bed tubular reactor: yields and structural analysis of bio-oil. Journal of Analytical and Applied Pyrolysis, 52133490165-2370

37. D. Radlein, 1999The production of Chemicals from Fast Pyrolysis Bio-oils. Fast Pyrolysis of Biomass: A handbook. CPL Press, Newbury.

38. F. Rick, U. Vix, 1991Product standards for pyrolysis products for use as fuel in industrial firing plant. in: Biomass pyrolysis liquids upgrading and utilization, Elsevier applied science, 177218

39. B. Valle, A. G. Gayubo, A. T. Aguayo, 2010Selective Production of Aromatics by Crude Bio-oil Valorization with a Nickel-Modified HZSM-5 Zeolite Catalyst. Energy & Fuels, 24206020700887-0624

40. R. H. Venderbosch, A. R. Ardiyanti, J. Wildschut, 2010Stabilization of biomass-derived pyrolysis oils. Journal of Chemical Technology and Biotechnology, 8556746860268-2575

41. S.Vitolo, P. Ghetti, 1994PhysicalAndCombustionCharacterization Of Pyrolytic Oils Derived From Biomass Material Upgraded By Catalytic-Hydrogenation. Fuel, 7311181018120016-2361

42. S. Vitolo, M. Seggiani, P. Frediani, 1999Catalytic upgrading of pyrolytic oils to fuel over different zeolites. Fuel, 7810114711590016-2361

43. B. M. Wagenaar, 1994The rotating cone reactor for rapid thermal solids processing, PhD, University of Twente.

44. Q. Wang, Q. Liu, B. He, 2008Experimental research on biomass flash pyrolysis for bio-oil in a fluidized bed reactor. Journal of Engineering Thermophysics, 298858880025-3231X.

45. S. R. Wang, Y. L. Gu, Q. Liu, 2009Separation of bio-oil by molecular distillation. Fuel Processing Technology, 9057387450378-3820

46. S. R. Wang, Z. Y. Luo, L. J. Dong, 2002Flash pyrolysis of biomass for bio-oil in a fluidized bed reactor. Acta Energiae Solaris Sinica, 11851880254-0096

47. J. Wildschut, F. H. Mahfud, R. H. Venderbosch, 2009Hydrotreatment of Fast Pyrolysis Oil Using Heterogeneous Noble-Metal Catalysts. Industrial & Engineering Chemistry Research, 482310324103340888-5885

48. M. M. Wright, J. A. Satrio, R. C. Brown, 2010Techno-Economic Analysis of Biomass Fast Pyrolysis to Transportation Fuels. NREL.

49. J. Yang, D. Blanchette, B. D. C. , C. Roy, 2001Progress in thermochemical biomass conversion.

50. Y. Yao, S. R. Wang, Z. Y. Luo, 2008Experimental research on catalytic hydrogenation of light fraction of bio-oil. Journal of Engineering Thermophysics, 2947157190025-3231X.

51. Q. Zhang, J. Chang, T. J. Wang, 2006Upgrading bio-oil over different solid catalysts. Energy & Fuels, 206271727200887-0624

Development of Eco-friendly Biodegradable Biolubricant Based on Jatropha Oil

M. Shahabuddin[1], M. Rahman[1], H.H. Masjuki[1], and M.A. Kalam[1]

[1]Centre for Energy Sciences, Faculty of Engineering, University of Malaya, Kuala Lumpur, Malaysia

INTRODUCTION

Various types of lubricants are available all over the world including mineral oils, synthetic oils, re-refined oils, and vegetable oils. Most of the lubricants which are available in the market are based on mineral oil derived from petroleum oil which are not adaptable with the environment because of its toxicity and non-biodegradability [1, 2]. Unknown petroleum reserve and the increasing consumption, which made concern to use petroleum based lubricant thus, to find the alternative lubricant to meet the future demand is an important issue [3]. Therefore, vegetable oil can be played a vital role to substitute the petroleum lubricant as it possesses numerous advantage over base lubricant like renewability, environmentally friendly, biodegradability, less toxicity and so on [4-8]. It has been reported that yearly 12 million tons of lubricants waste are released to the environment [9]. However, it is very difficult to dispose it safely for the mineral oil based lubricants due its toxic and non-biodegradable nature. To reduce the dependency on petroleum fuel, legislations have been passed to use certain percentage of biofuel in many countries, such initiative also required for lubricant as well [10]. Vegetable oils are mainly triglycerides which contain three hydroxyl groups and long chain unsaturated free fatty acids attached at the hydroxyl group by ester linkages acids favors triglycerides crystallization [11, 12]. The unsaturated free fatty acid which is defined as the ratio and position of carbon-carbon double bond, one two and three double bonds of carbon chain is named as a oleic, linoleic, and linolenic fatty acid components respectively [13]. The main limitations of vegetable oil are its poor low temperature behavior, oxidation and thermal stability and gumming effect [14, 15]. These stabilities and pour point behavior can be ameliorated by transesterification. Moreover the inferior flow property does not affect much in the tropical countries. Quinchia et al. [16] stated that, improving the potentiality of biolubricants some technical properties including available range of viscosities are need to improved. To do so, environmentally friendly viscosity modifier can be used. viscosity is the most important property for the lubricants since it determines the amount of friction that will be encountered between sliding surfaces and whether a thick enough film can be build up to avoid wear from solid-to-solid contact. Since little chance of viscosity with fluctuations in temperature is desirable to keep variations in friction at a minimum, fluid often are rated in terms of viscosity index. The less the viscosity is changed by temperature, the higher the viscosity index. Ethylene–vinyl acetate (EVA) and styrene–butadiene–styrene (SBS) copolymers were used to increase the viscosity range of high-oleic sunflower oil, in order to design new environmentally friendly lubricant formulations with increased viscosities. The maximum kinematic viscosities, at 40 and 100 °C, were increased up to around 150–250 cSt and 26–36 cSt, respectively [17].

Despite of having lot of advantages of biolubricant over petroleum based lubricant, the attempt to formulate the biolubricant and its applications are very few. Thus, in this article we sought to extend our investigation and to test the tribological characteristics and compatibility of non-edible Jatropha oil based biolubricant for the automotive application. The reason of selecting Jatropha oil as a base stock is it does not contend with the food and can be grown in marginal land.

EXPERIMENTAL

Lubricant Sample Preparation

There were six different types of lubricant sample were investigated in this study. The lubricant SAE 40 was used as a base lubricant and comparison purpose. Others samples were prepared by mixing of 10%, 20%, 30%, 40% and 50% Jatropha oil in SAE 40. The samples were mixed with the base lubricant by a homogeneous mixture machine.

Friction and Wear Evaluation

The apparatus used in the friction and wear testing process were Cygnus Friction and Wear Testing Machine which is connected with a personal computer (PC) with data acquisition system. It is a tri-pin-on-disc machine which is conducted by using three pins on a disc as testing specimens. Specifications of the Cygnus Test Machine are tabulated in Table 1. The block diagram of friction and wear testing are shown in Fig. 1. During the test the load of 30N and rotational speed of 2000 rpm were applied on pin.

Table 1: Specification of Cygnus wear testing machine

Parameter	Value
Test Disc Diameter	110.0 mm
Test Pin Diameter	6.0 mm
Test Disc Speed Range	25 to 3000 rpm
Motor	Tuscan; (2000 rpm, 1.5 kW)
Load Range	0 KG to 30 KG
Electrical Input	220 Volt AC 50 Hz

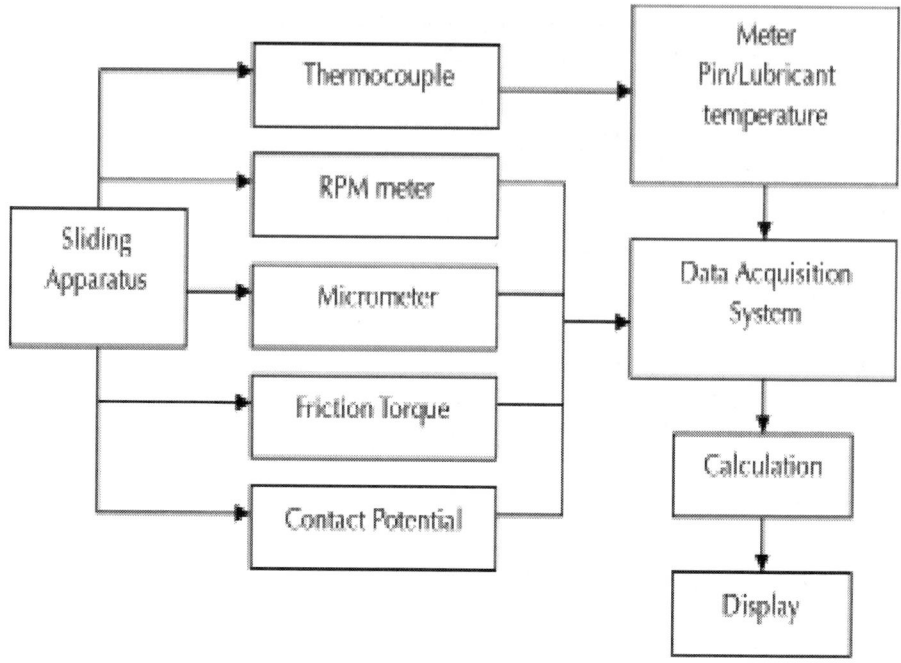

Figure 1: Block Diagrams of Friction and Wear Testing.

Preparation of the Specimen

The specimens were prepared from aluminum and cast iron material. Aluminum was used to build three pin and cast iron is used for disc specimen. The construction geometry and the dimension are shown in Fig. 2. Prior to conduct the test it was ensured that the surface of the specimens are cleaned properly i. e, free from dirt and debris. Alcohol was used for cleaning purpose.

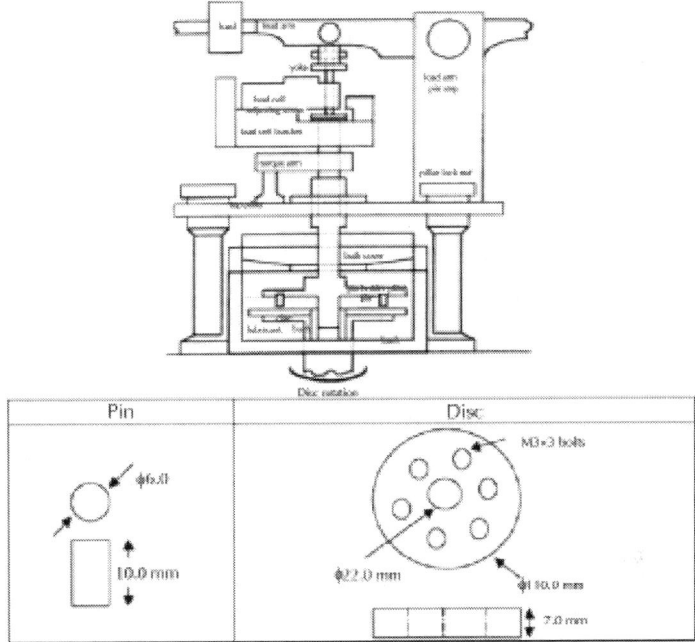

Figure 2: Schematic diagram of the experimental set up and dimensions geometry's of pins and disc specimen.

Lubricant Analyses

Multi element oil analyzer (MOA) was used to measure the wear elements in the lubricants by Atomic Emission Spectroscopy (AES). Whereas, for viscosity measurement the automatic Anton Paar viscosity meter was used with standard ASTM D 445. Viscosity was measured for both 40°C and 100°C controlled bath temperatures.

RESULTS AND DISCUSSION

Friction and Wear Characterization

Fig. 3 show the pins wear as a function of sliding time for various Jatropha oil blended biolubricants. At the operating condition of 2000 rpm and 30 N loads, the linear pin wear varied from 0.02 to 0.05 mm. It is observed that the maximum wear occurred in the beginning of the experiment using biolubricants. It can be seen form the Fig. 3, that the maximum wear was occurred for JBL40 while the minimum wear was observed for JBL10. The results can be attributed to the maximum ability of the JBL 10 biolubricant film to protect metal to metal contact and keep consistency throughout the operation time while this ability is least for JBL40. It can also be seen that the rate of wear throughout the time is almost identical for the biolubricants whereas, the reducing trend is observed for the base lubricant. At the beginning of the test, the wear rate was very fast for few minutes which are known running-in period. During this period, the asperities of the sliding surface are cut off and the contact area of the sliding surface grows to an equilibrium size. After certain period of time, equilibrium wear condition between pins and disc surface was established and thereby the wear rate became steady. It can be identified from the Fig. 3 that the biolubricants JBL 30, JBL 40 and JBL 50 showed high wear while base lubricant, JBL 10 and JBL 20 impart low pin wear and their values are nearly same with each other.

Fig. 4 sows the loos of material from the pin for different percentage of biolubricant samples. It seems quite clear that the loos of material from the pins are highest for 50% biolubricant and that is least for base lubricant. It can also be interpreted that the loos of material from JBL 10 is almost similar with base lubricant and this loos of material is increasing with increasing biolubricant percentages.

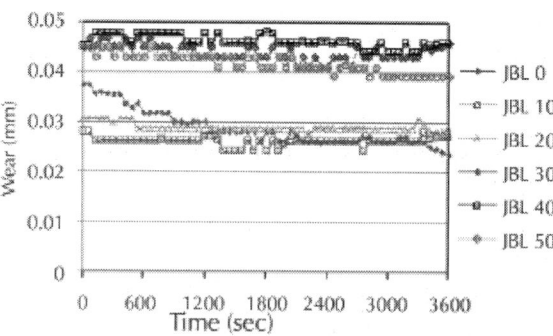

Figure 3: The linear pin wear as a function of sliding time for various Jatropha oil biolubricants.

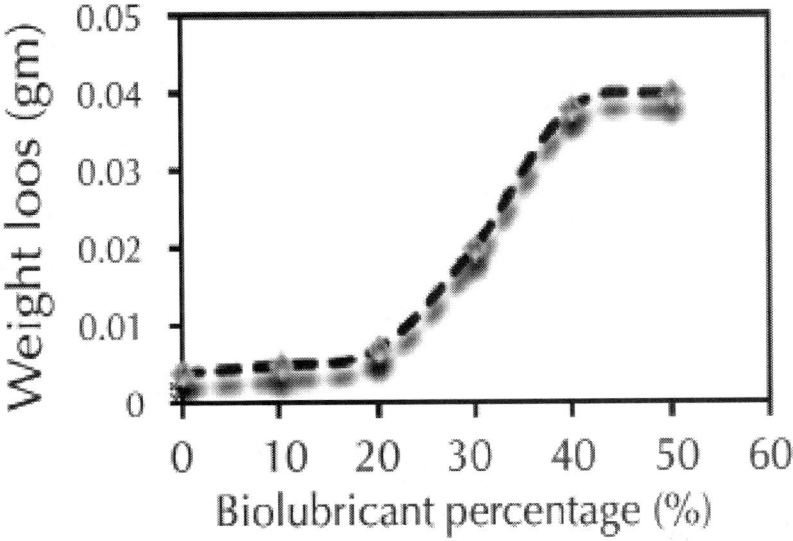

Figure 4: Loos of material form the pin for various biolubricant percentages.

Coefficient of Friction

Fig.5 shows the friction coefficient plotted against the sliding time for various Jatropha oil biolubricants. The results of the figure depict that the lubricant regime that occurred during the experiment were the boundary lubrication with the value of friction coefficient for boundary lubricant in the range of 0.001 to 0.2 except for 50% of Jatropha oil biolubricant. For JBL 0, it can be seen that the coefficient of friction is highest at the beginning and then it fell down rapidly and became least with compared to all tested samples after half of the operation time. The biolubricant percentage from 10 to 40% showed likely to be similar coefficient of friction (μ) which is almost 0.15. Whereas, the 50 % added Jatropha oil showed the coefficient of friction value of ~ 0.225 throughout the operation time. The fatty acid component of biolubricants formed multi and mono layer on the surface of the rubbing zone and make stable film to prevent the contact between the surfaces.

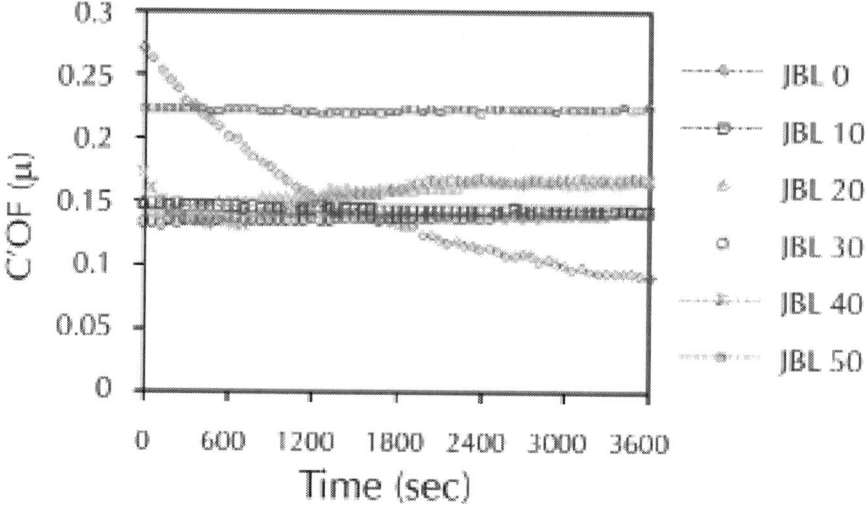

Figure 5: The Coefficient of friction as a function of sliding time for various Jatropha oil biolubricants.

Lubricants Temperature

Fig.6 shows the relationship of the averages oil temperature of varies percentage of Jatropha oil biolubricants with the sliding time. The rise of temperature during the running hour (1 h) for JBL 10 is least while the highest change is occurred for JBL 40 which is 11.77°c and 25.49°C respectively. The temperature rises of other samples are of 12.8°C, 18.65°C and 13. 66°C for 20% 30% and 50% Jatropha oil added biolubricants respectively. The results of the Fig. 6 show that the JBL 10 has the highest potentiality to retain its property without much changing its temperature. From the figure it can also be interpreted that up to 30 minutes rate of change of temperature is high while the changing rate is low for second half of the operation time. It can be explained that during second half of the operation time heat produced in the lubricant due friction and the heat dissipated to the outside is nearly equilibrium.

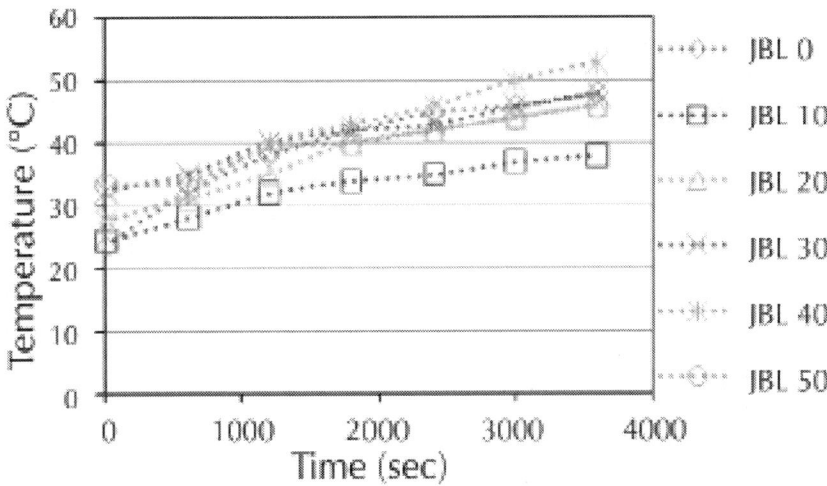

Figure 6: The Lubricant Temperature as a function of sliding time for various Jatropha oil biolubricants.

Viscosity

Viscosity is the measure of resistance to flow [18]. Table 2 shows the viscosity grade requirement for the lubricants set by International standard organization (ISO), while Fig. 7 shows the viscosity of tested different biolubricant samples. The comparison of the results of the Fig.7 with that of ISO grade illustrates that in case of 40°C, the biolubricants JBL 40 and JBL 50 did not meet the ISO VG100 requirement. On the other hand all other biolubricants meet the entire ISO grade requirement as well. It can also be noted that the viscosity of biolubricants are much higher than standard requirements

Table 2: ISO Viscosity grade requirement [19]

Kinematic viscosity	ISO VG32	ISO VG46	ISO VG68	ISO VG100
@ 40°C	"/>28.8	"/>41.4	"/>61.4	"/>90
@ 100°C	"/>4.1	"/>4.1	"/>4.1	"/>4.1

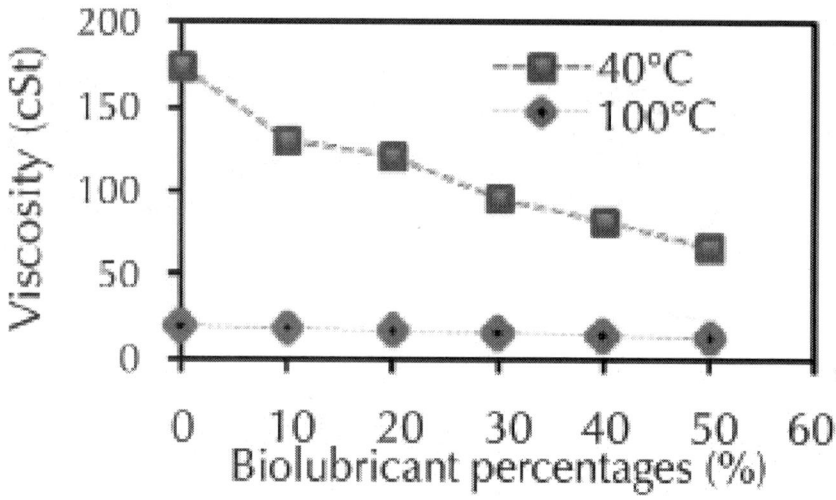

Figure 7: The viscosity of various percentages of biolubricants at 40°C and 100°C.

Elemental Analysis

The aim of the elemental analysis by using Multi Element Oil Analyzer (MOA) is to determine the kinds and amount of metal contain in the lubricating oil. Table 3 shows the elemental analysis of tested lubricant sample by using MOA before and after the test. From the Table 3, it can be noticed that the base lubricant contains higher Silver (Ag), Zinc (Zn), Phosphorus (P), Magnesium (Mg) and Boron (B) with in high percentage compared to other element while, in pure Jatropha oil, Calcium (Ca) and Silicon (Si) are the higher element compared with other element. Some of the elements are used as additive in the lubricant to ameliorate the lubricants tribological properties. From the results, increasing number of iron (Fe) and aluminum (Al) molecules are observed with increasing percentages of Jatropha oil in the base lubricants. The source of Fe and Al are mainly cast iron plate and aluminum plate. Due to lower hardness of the aluminum pin the extraction of aluminum molecule form the pin is much higher than cast iron plate. The changes of other elements were observed before and after the test. It is clear from the elemental analysis that, most of elements were decreased after the test, by oxidizing and the chemical interaction among the elements.

Table 3: Elemental analysis of tested lubricant sample

Parameters Test	JBL 0 Before	JBL 0 After	JBL 10 Before	JBL 10 After	JBL 20 Before	JBL 20 After	JBL 30 Before	JBL 30 After	JBL 40 Before	JBL 40 After	JBL 50 Before	JBL 50 After	Iatropha oil
Iron (Fe)	0.00	200	1.00	2.00	1.00	3.00	1.00	3.00	1.00	6.00	2.00	6.130	2
Aluminum (Al)	0.00	15.00	0.00	81.00	0.00	188.00	0.00	205	0.00	2110	0.00	76.00	0
Copper (Cu)	0	1.00	0.00	3.00	1.00	1.00	1.00	1.00	1.00	7.00	2.00	5.00	3
Lead (Pb)	3	4.00	400	5.00	200	4.00	3.00	4.00	3.00	3.00	3.00	2.00	0
Tin (SO)	0.00	0.00	0.00	0.00	0.00	0.00	0.00	0.00	1.00	0.00	2.00	2.00	4.5
Nickel (Ni)	2.00	2.00	3.00	3.00	1.00	3.00	3.00	3.00	3.00	3.00	2.00	2.00	1.5
Titanium (Ti)	0.00	0.00	1.00	1.00	0.00	1.00	0.00	1.00	0.00	0.00	1.00	1.00	1
Silver (Ag)	108	103	0.00	0.00	0.00	0.00	0.00	0.00	0.00	0.00	0.00	0.00	0
Molybdenum (Mo)	3.00	3.0	4.00	6.00	2.00	3.00	4.00	3.00	4.00	6.00	3.00	4.00	1.5
Zinc (Zn)	1000	771	903	716	1000	829.0	911.0	851.0	942.00	900.0	946.00	832.00	1
Phosphorus (P)	500.00	428	471	441	462.00	440.0	435.00	408.0	387.00	394.0	348.00	294.00	45
Calcium (Ca)	18.00	17.00	21.00	29.00	23.00	21.0	28.00	27.0	35.00	33.00	37.00	30.00	40
Magnesium (Mg)	748.00	637.0	572.	616.00	557.00	435.0	503.00	527.0	508.00	483.0	409.00	211.00	27

Silicon (Si)	5.00	4.00	6.00	10.00	6.00	15.0	8.00	12.0	9.00	13.0	14.00	7.00	16
Sodium (Na)	2.00	1.00	200	5.00	200	2.0	3.00	4.00	5.00	4.00	3.00	4.00	4
Boron (B)	60.00	54.00	52.00	58.00	52.00	28.0	52.00	32.0	44.00	44.0	40.00	21.00	05
Vanadium (V)	0.00	1.00	0.00	1.00	0.00	0.00	1.00	0.00	1.00	0.00	1.00	1.00	1

Surface Texture Analysis

There are various types of wear in the mechanical system, such that abrasive wear, adhesive wear, fatigue wear and corrosive wear. Since the lubricant regime occurred in this experiment was boundary lubrication thereby, abrasive wear, adhesive wear, fatigue wear and corrosive wear were observed in to the rubbing zone. All these wears mechanisms found in this experiments but the mostly the wear phenomenon were abrasive and adhesive wear. This is because of an existence of straight grooves in the direction of the sliding direction. These grooves exist because the asperities on the hard surface (disc) touched the soft surface (pins) and hade a close relationship with the thickness of lubrication film. The optical images of the tested cast iron plate using various types of biolubricants are shown in Fig. 8. Referring to the Fig. 8, it is found that the wear increases with increasing percentage of Jatropha oil in the biolubricants. Reduction of lubricant film thickness leads to the surfaces to come closer to each other and cause higher wear.

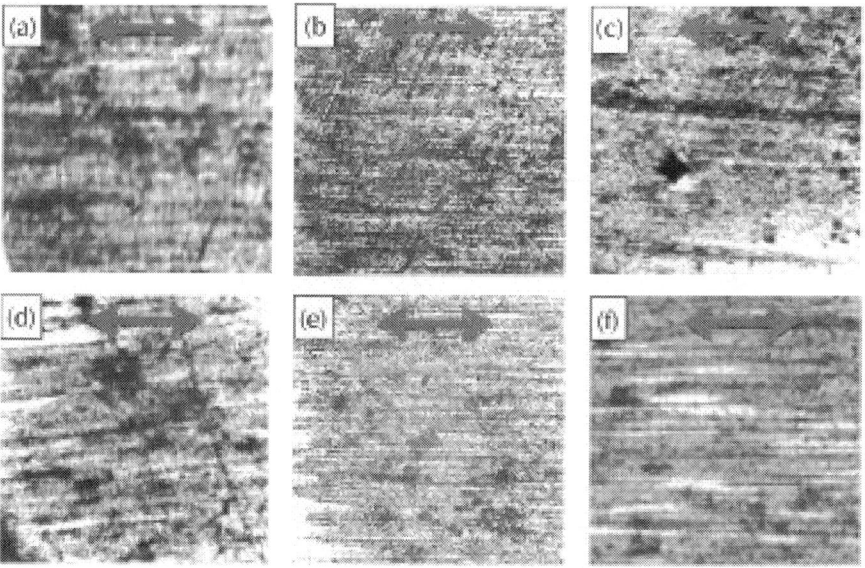

Figure 8: Optical image of the surface of the cast iron plate for different bio-lubricants (magnification 30 ×): (a): JBL 0, (b): JBL10, (c): JBL 20, (d): JBL 30, (e): JBL 40, (f): JBL 50.

CONCLUSIONS

Based on the experimental study the following conclusion can be drawn:

- M. Shahabuddin[1], M. Rahman[1], H.H. Masjuki[1], and M.A. Kalam[1]
- M. Shahabuddin[1], M. Rahman[1], H.H. Masjuki[1], and M.A. Kalam[1]
- The rates of wear for various percentage of biolubricant were different. Moreover the wear rate for 10% Jatropha added biolubricants were almost identical with base lubricant.
- Lower the resistance to wear, higher coefficient of friction.
- At the beginning of the test rate of wear as well as rise in temperature were high. With respect to wear rate and rise in temperature during entire operation time, the JBL 10 biolubricant showed best performance in terms of its ability to withstand its properties.
- From the elemental analysis of the biolubricants, it is found, Iron and Aluminum were increased after the test due to the loos of material from the pin and the disc while, some element like Phosphorus, Calcium and Magnesium were decreased by oxidizing and due to other chemical interaction.
- In terms of viscosity, almost all biolubricants met the ISO viscosity grade requirement whereas, 40% and 50% addition of Jatropha oil do not meet the ISO VG 100 requirement at 40°C.

According to the experimental result, it can be recommended that the addition of 10% Jatropha oil in the base lubricant is the optimum for the automotive application as it showed best overall performance in terms of wear, coefficient of friction, viscosity, rise in temperature etc.

ACKNOWLEDGEMENTS

The authors would like to acknowledge the Department of Mechanical Engineering, University of Malaya, Ministry of Higher Education (MOHE) of Malaysia for HIR grant (Grant No. UM.C/HIR/MOHE/ENG/07) and ERGS grant no ER022-2011A which made this study possible.

REFERENCES

1. N. Salih, J. Salimon, E. Yousif, Synthetic biolubricant basestocks based on environmentally friendly raw materials. Journal of King Saud University-Science 2011

2. A. Adhvaryu, Z. Liu, S. Erhan, Synthesis of novel alkoxylated triacylglycerols and their lubricant base oil properties. Industrial Crops and Products 200521113119

3. M. Shahabuddin, H. H. Masjuki, et. Kalam, al, Effect of Additive on Performance of C.I. Engine Fuelled with Bio Diesel. Energy Procedia 20121416241629

4. M. T. Siniawski, N. Saniei, B. Adhikari, L. A. Doezema, Influence of fatty acid composition on the tribological performance of two vegetable-based lubricants. Journal of Synthetic Lubrication 200724101110

5. Salunkhe DK. World oilseeds: chemistry, technology, andutilization. 1992

6. Hwang HS, Erhan SZ.Lubricant base stocks from modified soybean oil. AOCS Press: Champaign, IL; 2002

7. T. C. Ing, A. K. M. Rafiq, S. Syahrullail, Friction Characteristic of Jatropha Oil using Fourball Tribotester. In: Regional Tribology Conference- RTC2011. Langkawi, Kedah, Malaysia: 2011

8. M. Shahabuddin, M. A. Kalam, H. H. Masjuki, M. Mofijur, Tribological characteristics of amine phosphate and octylated/ butylated diphenylamine additives infused biolubricant. Energy Education Science and Technology Part A: Energy Science and Research 20123089102

9. G. E. Totten, S. R. Westbrook, R. J. Shah, Fuels and Lubricants Handbook: Technology,Properties, Performance, and Testing. 2003885909p.

10. Liaquat AM, Masjuki HH, Kalam MA et al.Application of blend fuels in a diesel engine. Energy Procedia 20121411241133

11. N. Jayadas, K. P. Nair, Coconut oil as base oil for industrial lubricants--evaluation and modification of thermal, oxidative and low temperature properties. Tribology international 200639873878

12. N. Fox, G. Stachowiak, Vegetable oil-based lubricants-a review of oxidation. Tribology international 20074010351046

13. C. Waleska, E. W. David, C. Kraipat, M. P. Joseph, The effect of chemical structure of base fluids on antiwear effectiveness of additives. Tribol. Int. 2005383216

14. N. Ponnekanti, S. Kaul, Development of ecofriendly/ biodegradable lubricants: An overview. 2012

15. M. Mofijur, H. H. Masjuki, et. Kalam, al, Palm Oil Methyl Ester and Its Emulsions Effect on Lubricant Performance and Engine Components Wear. Energy Procedia 20121417481753

16. L. Quinchia, M. Delgado, C. Valencia, et al. Viscosity modification of different vegetable oils with EVA copolymer for lubricant applications. Industrial Crops and Products 201032607612

17. L. Quinchia, M. Delgado, C. Valencia, et al. Viscosity modification of high-oleic sunflower oil with polymeric additives for the design of new biolubricant formulations. Environmental science & technology 20094320602065

18. M. Shahabuddin, Masjuki. H. H. Kalam, et al. An experimental investigation into biodiesel stability by means of oxidation and property determination. Energy 2012

19. Rudnick LR. Automotives Gear Lubricants, Synthetics, mineral oils, and bio-based lubricants: chemistry and technology. Taylor and Francis, Florida; 2006

Citations

CHAPTER 1

Haiyang Yu, Hui Guo, Youwei He, et al., "Numerical Well Testing Interpretation Model and Applications in Crossflow Double-Layer Reservoirs by Polymer Flooding," The Scientific World Journal, vol. 2014, Article ID 890874, 11 pages, 2014. doi:10.1155/2014/890874.

CHAPTER 2

Zhilin Liu, Lutao Liu, and Jun Zhang, "Signal Feature Extraction and Quantitative Evaluation of Metal Magnetic Memory Testing for Oil Well Casing Based on Data Preprocessing Technique," Abstract and Applied Analysis, vol. 2014, Article ID 902304, 9 pages, 2014. doi:10.1155/2014/902304.

CHAPTER 3

ia Zhichun, Li Daolun, Yang Jinghai, Xue Zhenggang, and Lu Detang, "Numerical Well Test Analysis for Polymer Flooding considering the Non-Newtonian Behavior," Journal of Chemistry, Article ID 107625, in press.

CHAPTER 4

Alexandre Andrade Cerqueira and Monica Regina da Costa Marques (2012). Electrolytic Treatment of Wastewater in the Oil Industry, New Technologies in the Oil and Gas Industry, Dr. Jorge Salgado Gomes (Ed.), ISBN: 978-953-51-0825-2, InTech, DOI: 10.5772/50712.

CHAPTER 5

Adesina Fadairo, Churchill Ako, Abiodun Adeyemi, Anthony Ameloko, and Olugbenga Falode, Novel Formulation of Environmentally Friendly Oil Based Drilling Mud, doi: 10.5772/51236.

CHAPTER 6

Shurong Wang (2013). High-Efficiency Separation of Bio-Oil, Biomass Now - Sustainable Growth and Use, Dr. Miodrag Darko Matovic (Ed.), ISBN: 978-953-51-1105-4, InTech, DOI: 10.5772/51423.

CHAPTER 7

M. Shahabuddin, M. Rahman, H.H. Masjuki, and M.A. Kalam (2013). Development of Eco-Friendly Biodegradable Biolubricant Based on Jatropha Oil, Tribology in Engineering, Dr. Hasim Pihtili (Ed.), ISBN: 978-953-51-1126-9, InTech, DOI: 10.5772/51376.

Index